THE ADVANCED PISTOL MARKSMANSHIP INSTRUCTOR'S MANUAL

UNITED STATES ARMY MARKSMANSHIP TRAINING UNIT

Fredonia Books
Amsterdam, The Netherlands

The Advanced Pistol Marksmanship Instructors'
Manual

by
United States Army

ISBN: 1-4101-0033-2

Copyright © 2002 by Fredonia Books

Reprinted from the original edition

Fredonia Books
Amsterdam, The Netherlands
http://www.fredoniabooks.com

All rights reserved, including the right to reproduce
this book, or portions thereof, in any form.

EFFECTIVE INSTRUCTION IS AN ART. LIKE ANY ART IT PERMITS THE EXERCISE OF INITIATIVE AND INGENUITY. THE PRINCIPLES AND TECHNIQUES OF INSTRUCTION USED IN THIS MARKSMANSHIP TRAINING PROGRAM HAVE PROVED EFFECTIVE FOR TRAINING MILITARY PERSONNEL TO BECOME INFORMATIVE AND COMPETENT TEACHERS. USED INTELLIGENTLY AND SKILLFULLY, THESE PRINCIPLES WILL ACHIEVE THE OBJECTIVE: OPTIMUM UNDERSTANDING IN THE MIND OF THE STUDENT SHOOTER.

UNITED STATES ARMY MARKSMANSHIP TRAINING UNIT

PISTOL   DIVISION

MISSION: To produce pistol teams of the highest possible standard of proficiency capable of representing the United States Army by winning in national and international competitive events. Further, to establish doctrine and develop techniques in pistol shooting and coaching which can be utilized at all echelons; to distribute this knowledge by publication of training manuals and conduct training programs at various levels of command.

THE ADVANCED PISTOL MARKSMANSHIP INSTRUCTOR'S MANUAL
VOLUME II
TABLE OF CONTENTS

PAGE

FOREWORD ... 1

SECTION I. Technique of Pistol Marksmanship Instruction ... 2

    CHAPTER I. The Learning Process ... 5
    CHAPTER II. Methods of Instruction ... 14
    CHAPTER III. Lesson Planning ... 37
    CHAPTER IV. Effective Speaking ... 57
    CHAPTER V. Control of Interest ... 69
    CHAPTER VI. Management of Instruction ... 78
    CHAPTER VII. Training Aids ... 87
    CHAPTER VIII. Review of Course of Instruction ... 97
    CHAPTER IX. Examination, Critique, and Panel Discussion ... 100

SECTION II. Program of Instruction for Advanced Pistol Marksmanship ... 109

    CHAPTER X. Training Standards and Scheduling of Courses of Instruction ... 110
    CHAPTER XI. Lesson Plans for Pistol Marksmanship Course ... 123
        A. Attaining a Minimum Arc of Movement ... 125
        B. Sight Alignment ... 131
        C. Trigger Control ... 138
        D. Establishing a System ... 142
        E. Technique of Slow Fire ... 151
        F. Technique of Sustained Fire ... 159
        G. Mental Discipline ... 165
        H. Attributes, Responsibilities and Duties of a Pistol Team Coach ... 175
        I. Technique of Coaching a Pistol Team ... 180
        J. Evaluation of the Pistol Team Shooter ... *
        K. Physical Conditioning ... 189
        L. Diet of the Competitive Pistol Shooter ... 194
        M. Effects of Alcohol, Coffee, Tobacco and Drugs ... 199
        N. Pistol Team Organization and Administration Including
        O. Procurement, Maintenance, and Security of Match Weapons, Equipment, and Ammunition ... 213
        P. NRA Pistol Match Rules Including NRA Pistol Range Procedure and Safety Rules ... 217
        Q. National Trophy Pistol Individual and Team Match Rules Including Requirements for Earning a Distinguished Pistol Shot Badge ... *
        R. Review of Course of Instruction ... 204
        S. Examination and Critique with Panel Discussion ... 211

*To be published at a later date.

		PAGE
SECTION III. Administration		224
CHAPTER XII.	Pistol Team Organization and Administration.	226
CHAPTER XIII.	Procurement, Maintenance, and Security of Match Weapons, Equipment, and Ammunition	242
SECTION IV. Competitive Regulations.		246
CHAPTER XIV.	NRA Pistol Match Rules	248
CHAPTER XV.	NRA Pistol Range Procedure and Safety Rules	268
CHAPTER XVI.	National Trophy Pistol Individual and Team Match Rules	272
CHAPTER XVII.	Requirements for Earning a Distinguished Pistol Shot Badge	281
INDEX		283
GLOSSARY OF INSTRUCTIONAL TERMS		105

FOREWORD

Acquiring the ability to accurately shoot a pistol is no simple matter. One should not assume that the art of advanced pistol marksmanship is fully realized immediately upon reading a training manual on the fundamentals and techniques of pistol shooting; nor completely understood after having received a few hours of advice and instruction from a qualified coach or expert shooter.

To become a top pistol shooter and able to produce consistently high scores, one must learn to perform all the fundamentals of shooting, acquire certain definite habits, achieve flawless coordination - and above all - have a capacity for the intense concentration essential to exercising a high degree of mental control. For this, one must train.

It is difficult, if not impossible, to establish a universal system of training in pistol marksmanship which will cover all cases - one that can be adapted to each shooter's technique or special need. Training pistol shooters requires an individual approach. There is no single pattern or system for the organization of training that will entirely meet the individual requirements of all competitors, nor can one ever be expected. This is the reason knowledgeable coaches take the pecularities of an individual - such as experience, degree of preparation and fitness, and other items into account. They use these, and by relying on their own experience, devise a training program which allows special consideration for each shooter's capabilities. In spite of the difference in details, technique or method, there is much that is common to the training of advanced marksmen, that - in the opinion of leading coaches and shooters, applies to everyone, without exception. This manual endeavors to present, in a detailed, comprehensive manner, these universal applications.

Advanced pistol marksmanship training must of necessity, avoid the involved and exceedingly complex because it is an activity whose participants form a great cross-section of our national life, and the average citizen is its greatest asset.

We are grateful to the United States Army Infantry School for their counsel, advice, and cooperation in the preparation of those portions of this volume which deal with techniques of instruction. Many of their proven instructional techniques have been adapted to pistol marksmanship instruction.

Your constructive comments are invited; they should be addressed to the:

Commanding Officer
USAMTU
Fort Benning, Georgia

JOS. J. PEOT
Colonel, Sig C
Commanding

SECTION ONE

TECHNIQUE OF PISTOL MARKSMANSHIP
INSTRUCTION

THE SUCCESSFUL INSTRUCTOR

Instructing involves presenting ideas, appreciations, understanding and doctrine to students in a concentrated marksmanship training schedule. To carry out this assignment successfully, the instructor must demonstrate certain positive qualities. For many years instructors have studied the problem of what is required to teach successfully. Many of the necessary characteristics are those which make a successful marksman. The instructional traits listed approximately in the order of importance are: knowledge of subject; demonstrated ability to communicate ideas; ability to organize lesson materials; knowledge of instructional methods and skill in techniques of presentation and an ability to grow professionally. Instructors must also have a favorable attitude towards teaching and exercise a fair degree of personal imagination.

The marksmanship instructor must have a deep, rich knowledge of his subject. An outstanding instructor does not parrot other instructor's examples and phrases inherited from the vault file. Rather, he constantly searches for current, meaningful examples or appropriate events to support his teaching points. The doctrine or teaching points may remain the same but the presentation has been personalized by his own research. So rich is his background that he can use a variety of presentation methods to maintain student interest. No two of his presentations of the same problem are identical even though the teaching points may be the same. The ebb and flow of student questions and discussion condition each problem presentation so that there is always something new and stimulating. He is aware of correct teaching methods, skills and techniques and flexible enough to adapt them to each specific teaching situation.

A good instructor realizes that securing rapport or a sense of close communication between himself and the student body is difficult. Nevertheless, he is quick to appreciate the attitude of his class and he adapts to the student point of view in presenting his material. It is this sensitivity to student reaction which makes him a successful instructor. He quickly realizes when his class is having difficulty in understanding the subject matter and attempts to clear up this difficulty by encouraging student participation through discussion. He is aware of the import and significance of students' questions and skillful at answering them and at summarizing student comments in language understandable to the class. By proper handling of student questions, comments, and responses he creates a receptive rapport within the class which facilitates learning. His thinking and expression are in terms of ideas, not words.

The lesson plan of a capable instructor reflects a good sense of organization. The instructor has carefully researched his material and outlined, planned and rehearsed his instructional problem. He has timed his presentation and every orderly element of it is an integral and necessary part. Since his students are working within a tight training schedule, it is imperative that his presentation start and stop on time. Yet his primary concern is to ensure student learning; hence, he must allow sufficient time for his students to understand the teaching points and to work the practice exercises which were designed to reinforce learning or to practice the application of the principles learned.

An instructor must display a great deal of personal force. Sometimes he may be presenting his instructional problem out-of-doors--on the range, on the field, or in the classroom. He is convinced of the truth and significance of what he is saying and his task is to persuade and convince his students. As the principal instructor, he is the acknowledged authority on the subject and he takes his stand firmly so there is no doubt left in the student's mind.

Although his diction and vocabulary may not be so polished as that of a network announcer or university professor, he nevertheless uses acceptable grammar and has the vocabulary and fluency of one man communicating to another. He can think on his feet and give competent answers to student's questions. He gains the respect and wholehearted cooperation of his class because of his sincerity, his sense of humor, and his force.

The instructor is sold on the worth and value of what he is presenting. His interest in teaching and his interest in his students and the subject matter is contagious. He is able to project this enthusiasm even to large classes of two hundred students so that they too take an active enthusiastic part in the learning process and involve themselves directly into the problem. It is this active student involvement in the problem and the give and take of thought-stimulating questions and ideas which is characterized by the team instructional conference.

Part of his task is to help train new instructors. While these beginners are attending instructor training he may be designated as the student's sponsor. He cheerfully supplies all necessary administrative and logistical support. He rehearses with him conscientiously. He is careful not to dominate by insisting on parroting his words or mimicking his actions. He knows that instructor individuality is essential for effective teaching.

The instructor's appearance plays a key part in his presentation. His voice must be firm, positive and convincing. His manner should be poised, friendly and personal.

The Pistol Marksmanship Instructor.

Chapter I

THE LEARNING PROCESS

A. GENERAL

1. INTRODUCTION. How do students learn? The answer to this question has baffled philosophers for centuries. Psychologists and educators have researched the problem. Yet there is no universally accepted solution. Basically, the learner is exposed to a new situation (or problem) and he reacts or responds to it. When he responds in a way that is satisfactory to him personally or to society generally, we say that he responds correctly. Thus learning can be reduced to reacting to situations. As an instructor, you are interested in what you can do to assist the student to learn correctly, quickly, and accurately, and to learn thoroughly so he retains what he learns. Therefore, the emphasis in this chapter will be on controlled learning situations - those which occur within a classroom or training area.

B. STEPS IN LEARNING

1. LEARNING STEPS. We speak of learning, not learn. By using the "ing" form of the word we imply an action that takes place over a period of time, not something that occurs once and is complete and finished. In itself, learning is a continuing process of improvement and increasing proficiency. Visualize learning as follows:

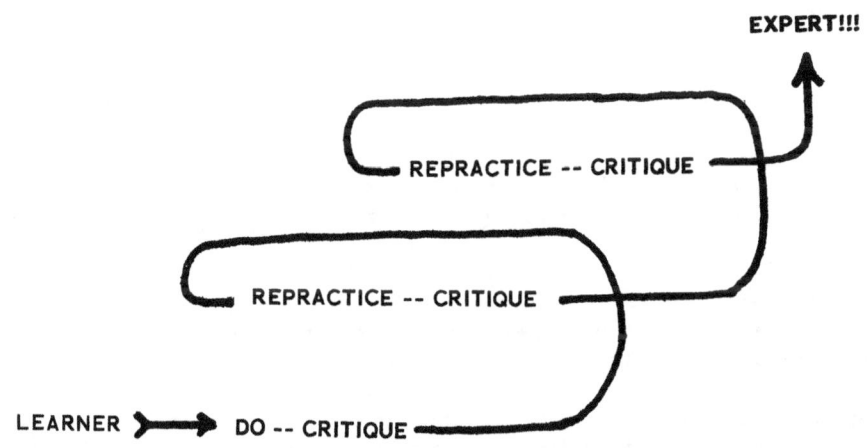

Figure 1-1. Spirals of Learning.

Learning is thus a process of continuing improvement with ascending spirals of practice, critique of practice, repractice, critique of repractice. With each new spiral there is an increase in understanding or skill as the learner strives to progress from novice to expert. Basically, there are four steps in the learning process:

| INTAKE | REACTION | CRITIQUE | REPRACTICE |
| (Observe, Listen) | (Do, Practice) | | |

Step 1 - The intake step - whenever the learner is exposed to the situation or problem. He uses his senses to observe, listen, and gather data.

Step 2 - The reaction step - whenever the learner reacts or responds to the situation by doing something. Frequently, this doing is practice of the material to be learned - whether this be a manual skill or a mental or problem-solving skill.

Step 3 - The critique or evaluation step. Such a critique may be personal evaluation of performances or it may be instructor evaluation of student performance.

Step 4 - The repractice step in the light of improvements suggested by the critique step.

NOTE: These steps are repeated as the student gradually improves or progresses toward qualified or expert proficiency.

C. TIPS TO ASSIST LEARNING

1. SUGGESTIONS FOR STEP 1, "THE INTAKE STEP".

a. During the intake step, the learner must understand what he is to learn and how he is to perform. As instructor, you must supply the purpose or motive for learning the material. You must show that your lesson will satisfy the student's need. Be sure that you understand why this lesson is necessary and that you have answered the students challenge of "why should I learn this?" Learning begins with a clear-cut purpose and a driving need.

b. Before the intake step can be effective, your students must be carefully prepared. They must be set or ready to learn. You do this by gaining and focusing attention in the lesson introduction. Next, you control interest during the intake step so students can get the most out of their observation and listening. With student interest high, their enthusiasm to learn will be high. Therefore, they will put forth more effort and drive to concentrate. Learning comes easier with concentration.

c. Use the "multiple-sense approach". Remember, students learn with their whole being. Insure that in your presentation you appeal to as many senses as possible. Stimulate sight, sound, physical doing, and mental activity by demonstrations, skits, training aids, questions, discussion, practice, and problem-solving exercises. The more senses you stimulate, the better chance for learning. The more vivid impression you make, the greater the learning imprint.

d. Use a variety of techniques to appeal to different groups in the class. One group will learn as a result of your initial statement and supporting material. Some will learn as a result of your rephrasing of the teaching point or additional supporting material. Others will learn by a student's answer to your question. Still other students will learn after you have shown a chart, given an example, or personal experience. Remember, among your students there are some who learn best by visualizing, some by listening, or some by discussing.

e. Supply the big picture or overview to help students acquire insight into the problem to be learned. Use the whole-part-whole technique.

(1) In lesson planning you do this by giving the big picture in the scope and motivation of the introduction, then you develop the "parts" as you present the different teaching points. Once again you present the "whole" as you wrap up the lesson with a comprehensive summary.

(2) Show them the perfection they are striving to reach as you demonstrate expert rifle disassembly, skilled marksman, or excellent examples of good teaching or public speaking. In this way the student sees what he is striving to equal and he better understands how these smaller learning units fit into the big picture of the finished learning product.

2. SUGGESTIONS FOR STEP 2, THE REACTION AND DOING STEP.

a. As the instructor you know that learning is not complete until the student practices what he has learned. He learns by doing. Your job is to supply meaningful practice - training the student under simulated conditions with practice or problem exercises designed to teach manual or mental skills clearly and effectively.

b. Divide the practice materials into small logical learning units or exercises. One of the few confirmed findings in educational psychology is this: FREQUENT SHORT PRACTICES ARE MOST EFFECTIVE FOR RAPID LEARNING AND LONGER RETENTION.

c. Don't introduce similar materials simultaneously that might confuse students. Students frequently learn and remember through association. They learn new things by relating them to things previously learned. Build up from known to the unknown (from that already learned to that which is to be learned). To study the nomenclature of two similar weapons simultaneously will confuse the average student.

d. Watch for learning plateaus. These are leveling-off places in the student's learning curve. At that time students are making no apparent improvement. This is normal for all students when learning a new skill. Explain this temporary stop over to the student so he does not get discouraged. It is a good time to schedule review, examination, or critique sessions. Here are examples of how students normally learn easy and difficult materials.

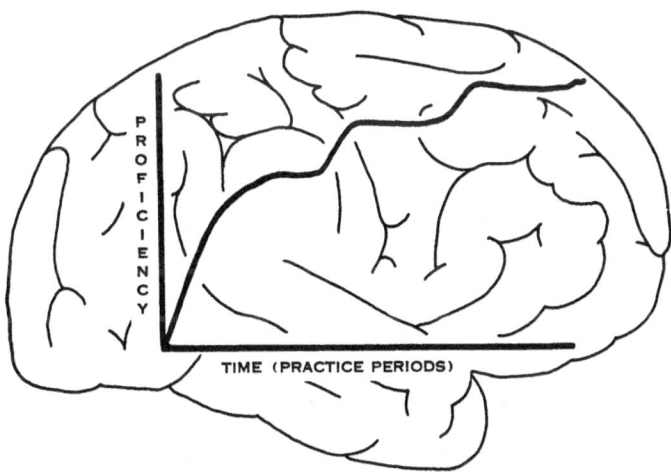

Figure 1-2. Normal Learning Curve - Easy Material.

NOTE: With easy materials there is a sharp initial spurt followed by learning plateaus and gradual slow improvement in proficiency. With learning most motor skills such as golfing, bowling, and swimming, you make good progress initially then it becomes increasingly difficult to reach the expert level.

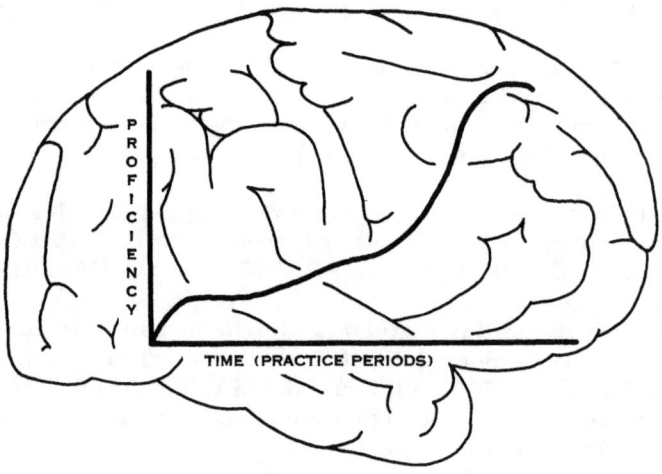

Figure 1-3. Normal Learning Curve - Difficult Material.

NOTE: With difficult material there is very slow progress until the fundamental concepts or basic principles are thoroughly understood (the first long plateau) then progress is faster toward the individual maximum personal level of proficiency.

3. <u>SUGGESTIONS DURING STEP 3, THE CRITIQUE STEP.</u>

    a. The critique or evaluation step may be self-evaluation of performance or instructor critique of student performance. Here are suggestions for the latter:

    b. Praise correct student performance as soon as possible. In this way the student will know what he is doing correctly. Encouragement is necessary early in student performance for every individual is apprehensive when he first practices. After all, he has seen expert demonstrations to set standards of performances and despairs of achieving such excellence.

    c. Point out only the gross errors at first. You must do more than this however. Show students exactly what they did wrong and try to explain why they probably performed incorrectly. Emphasize that such errors are normal with beginners.

    d. If the same student errors are repeated frequently, if they are dangerous errors, or when they are caused by inattention, take time to emphasize to the student that he must eliminate such errors for they will not be tolerated long.

    e. Test student performance frequently. Do this both for your benefit as the instructor and for the student's benefit. Both members of the learning team must know the status of student progress. As the student progresses, make your critiques more detailed and severe as you strive to smooth out the rough edges of performance.

4. <u>SUGGESTIONS DURING STEP 4, REPRACTICE STEP.</u>

    a. Topflight athletes agree that the three cardinal rules for expert proficiency are: practice, repractice, repractice. However, this repractice should be supervised, not blind repractice. With blind repractice, the student only ingrains the same wrong habits. Supervised repractice following a careful critique helps to set the correct way of performing and gradually eliminates errors.

b. At some stage in practice and repractice, under the eye of the professional, student finally acquires "insight" into the correct way of performing. In popular language we say he "gets the hang of it." Once the student realizes how to perform correctly and the common pitfalls to avoid, he can practice with only occasional supervision.

D. CONTROL OF LEARNING

1. LEARNING "CONTROL MEASURES".

a. A good instructor has various control measures to insure the success of his teaching mission. Many of these control measures for learning have been discussed in detail in other portions of this handbook; they are mentioned briefly here to show how they affect learning. Each or all of them have a significant impact on learning. Used correctly they can "make" the learning situation successful; neglected or misused they can "break" the learning situation.

b. Motivation - the student must have a need or motive for learning material; otherwise, he quickly forgets. For example, students who learned meaningless or nonsense words or phrases forgot them as quickly as they had learned them because they had no reason to retain them. When the student knows the purpose for learning material he learns it in order to apply such learning. This conditions the way in which he learns the material. Competition is one form of motivation. Individuals like to excel; Americans enjoy competition. Competition adds interest and encourages wholehearted participation. It can also provide valuable training in cooperation as long as you don't let competition get out of hand.

c. Interest - Motivation and interest are closely related. Since learning requires student action and concentration, you, as the instructor, must anticipate lack of interest, motivate and periodically remotivate, create the best teaching conditions and establish a favorable atmosphere for learning.

d. Action - Remember, learning is an active process, it doesn't occur by passive sitting and absorption. Students must perform, question, restate, analyze, argue, discuss, explain, and elaborate. Provide them with opportunities for such student action in the lessons you prepare. Get students to participate and react to the learning materials.

e. Logical Order - The order in which materials are presented will frequently affect learning success. As the instructor you guide the learning process. Arrange teaching materials in the simplest and most logical sequence. Make proper transitions from Step 1 to Step 2 so that the student understands how these steps relate or tie together. Some common sequences are:

(1) Chronological or orderly in time. With learning which involves procedures or steps be sure that you arrange them in the time order in which they occur.

(2) Whole-part-whole. This is the developmental order that psychologists claim is best to learn new materials. Give the big picture, then the details and then restate or summarize with the overview or big picture.

(3) Simple to complex. Go from the less difficult materials and build up to the complex. When students grasp fundamental ideas they can more readily apply them in advanced or complex situations.

(4) From known to the unknown. By this order you begin with concepts or principles already learned by the student. By comparison or analogy he learns how the new (unknown) material is similar to something he already knows. Tie in the new concept with past or everyday experiences.

(5) Orderly in space - Develop your explanations of objects so that you proceed from the left to the right, from top to bottom or from outside to inside. This will make student learning easier.

f. <u>Repetition</u>. Have the student repeat the learned action again and again. This helps to set a mental and muscular pattern. Tell him to make a conscious effort to remember as he performs. Use the principle of distributed practice - that is to space student application into frequent short practices so that fatigue does not waste training time. Don't interrupt. Be sure the student completes the action even though he is making mistakes and then critique. If you interrupt as you see errors he never has a chance to practice the action as a complete unit.

g. <u>Critique or Evaluation</u>. Evaluate progress periodically. This is important for the learner puts forth more effort if he knows he will be tested or checked regularly. Avoid general comments, be specific in your critique.

h. <u>Emotion</u>. Watch for strong feelings or emotions during performance and critique. Fear, anger, resentment, discouragement, and worry impede the learning process.

i. <u>Individual Differences</u>. Students are individuals. The nature and degree of individual differences in learning ability is important to the instructor. Class instruction is based on the assumption that people are enough alike that learning groups can be organized. Within these groups, however, you will find large differences in individual rate of learning. The fact that all individuals go through the same steps in learning does not mean that they all go through at the same speed, for the same reasons, with the same emotional reaction, or with the same net results.

Students differ in:

(1) Intelligence. Differences in individual ability are recognized in the use of aptitude area scores for classification and assignment in the Army. Wide differences in capacity for learning various types of performance will be found even in small groups.

(2) Level of Aspiration. Every student has his own level of aspiration. Some students set this so high that they are disappointed with their successes. They feel that they should be in the top ten of the class and if they place lower they are extremely disappointed with their achievement. Many students, on the other hand, set their goals so low that they are not motivated to do a creditable job. The goal-setting behavior of your students is influenced by their past experiences of success and failure. All students, to be learners, need success experiences commensurate with their abilities.

(3) Emotions. Watch out for strong feelings or emotions such as fear, anger, resentment, discouragement or worry for they will impede learning. Some students, the highly anxious group, are the worriers. They constantly worry about how they are making out in their studies. Experiments have demonstrated that this group often does more poorly on complex learning tasks than the less anxious group. It seems that the pressure on them to do better actually impedes their performance. When some pressure is put on the less anxious group, those who don't worry much about their school work, it spurs them to improve. Even after a task is learned, failure or the threat of failure may produce frustration and ruin a performance. Look at the shooters who "blow up" during stiff competition. Stress can produce varying results on students. Some do better under stress because they are more highly motivated. Others fold up under stress. Anxiety does affect learning.

(4) Past Experience. Relearning is easier than mastering new material. Some times a student will learn a skill in a tenth of the time it takes another student of equal ability to learn the same skill. The first student is drawing on past learning while the second is

encountering new material. This can occur even though the student who is learning it for the first time has more responses available, can make a better analysis of required responses, and has a better expectation of the learning goal.

E. PRODUCTS OF LEARNING

1. TRAINING GOALS. There are five common learning goals. The products of learning are skills, facts, concepts, preferences, and critical thinking ability.

a. Skills. Skills are acquired by drill. In firing a weapon, fingers and hands learn to work together by actually performing the act over and over until it becomes established as a pattern. Most skills require coordination, which means that many muscles must work together to produce the desired action. Repeated practice is necessary to achieve cooperation from several muscles - some contracting and some relaxing - to produce the skill pattern.

b. Facts. Facts are names, relationships, dates, and laws. They are often based on mental associations. These associations are fixed by meaningful repetition; the product is memory. Examples of such facts are: the elevation and windage rule, the cycle of operation.

c. Concepts or Principles. Concepts differ from facts primarily in depth of meaning that a student attaches to a word or other symbol. Concepts are usually the result of combining many basic facts together.

d. Preferences or Tastes. Preferences may be the result of both emotional and logical thinking. Usually they are the by-products of experience from other learning situations. A certain preferred way of acting brings satisfaction. Graduates of a technical or scientific school have preferences which differ markedly from those of the graduates of a liberal arts school. The high power shooter's preferences and tastes vary from those of the small bore or pistol shooter.

e. Critical Thinking Ability. Critical thinking involves manipulation of concepts and principles to arrive at logical conclusions. It has two phases: analysis, to dissect the problem into its elements and synthesis, to build up satisfactory solutions. It includes identifying basic issues, classifying evidence, and drawing sound deductions from data. Analytical thinking, or problem-solving ability, is learned by witnessing demonstrations of problem-solving techniques and by supervised practice in these techniques. Critical thinking in a specific area is learned by practice in evaluating concepts and weighing values in that area. Learning is fixed by experience in analyzing data, drawing conclusions, and determining appropriate courses of action.

F. HOW YOU REMEMBER

1. TYPES OF REMEMBERING. Psychologists who study learning recognize three types of remembering: recollection, recall and recognition.

a. Recollection. When you recollect, you reestablish an earlier personal experience on the basis of a partial clue. For example, the odor of a certain flower may bring back the memory of your first high school date who wore that flower to the school dance. Sometimes you can recollect the orchestra, decorations, names of school chums, where you went after the dance, etc. The details may or may not be complete. Recollections are distinguished from other kinds of remembering by reconstruction of a past occasion.

b. Recall. Recall is the human ability to perform some activity in the present which is based on past learning. You jump into water and swim although you may not have done so since childhood. You still remember the basic skills necessary. You can recite a poem you learned in grade school. Recall differs from recollection in that you cannot even remember the circumstances under which you learned to do this thing - all you can do is perform.

c. _Recognition_. In recognition, you recognize someone or something as familiar. It may be the way a person walks, the pitch of his voice, his accent, or it may be an odor or a street scene or river. You had experiences similar to this: you are walking down a street of a strange town and say to yourself, "I have been here before. There will be a drugstore on the corner and the theater will be down there." In this case, our memory is generalizing from similar street scenes and signalling faulty recognition. The present scene, though actually strange seems familiar to us.

2. THE CURVE OF REMEMBERING. The curve of remembering usually follows this pattern (See Figure 1-4). There is a rapid forgetting at first, followed by a leveling off or slowly decreasing losses. Frequently there is a slight gain after a short period before we begin to forget. We may lose as high as 50% of the material in this initial early period.

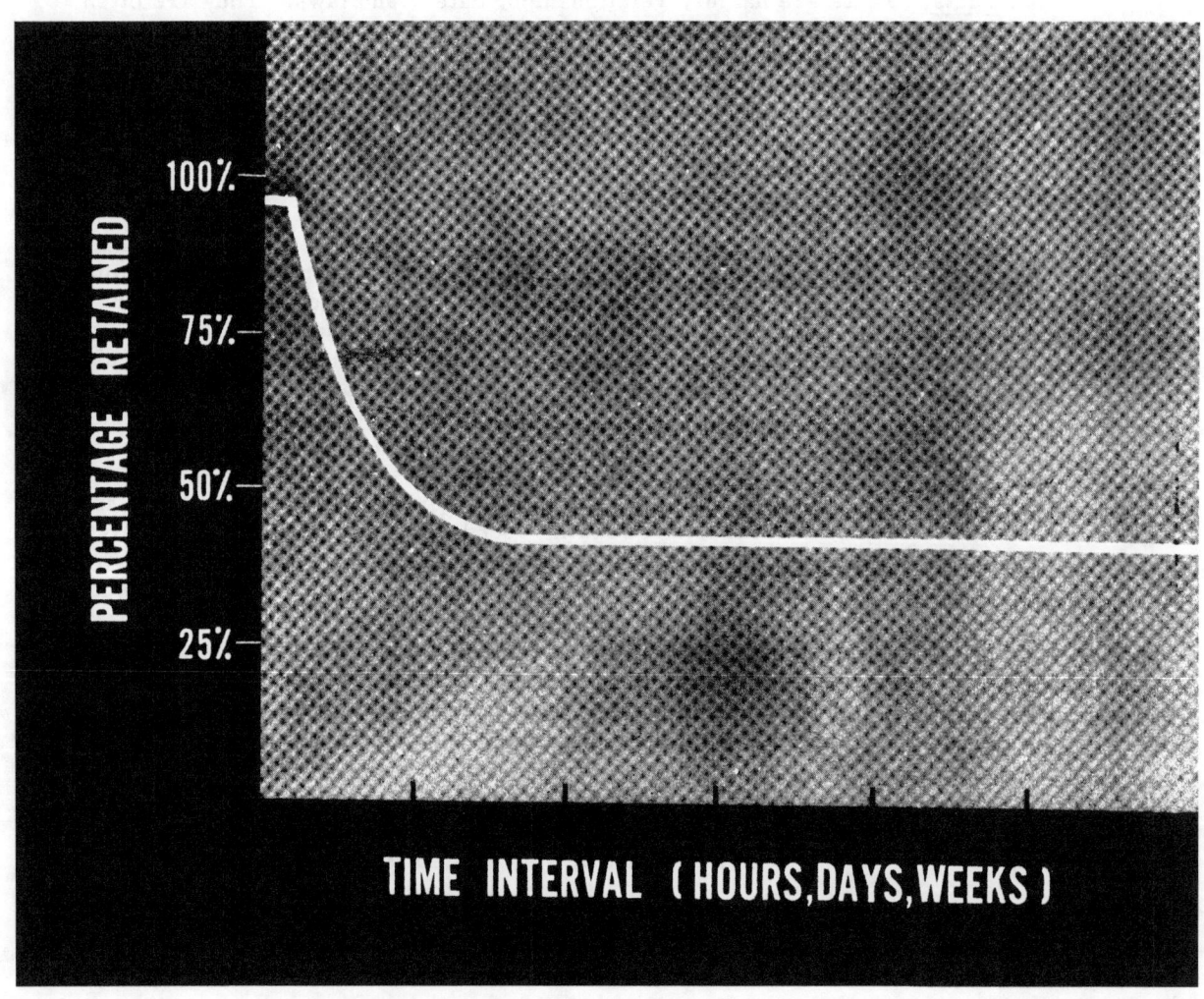

Figure 1-4. How You Forget.

G. WHY YOU FORGET

1. FORGETTING. You forget because of:

a. <u>Disuse.</u> The first cause for forgetting is the decay of the skill or memory pattern through lack of use. Although nothing is lost to the mind the student cannot call it back.

b. <u>Active Desire to Forget</u>. Many things you forget because your mind wants you to. Emotions enter into the picture to help you repress unpleasant or embarrassing experiences Experiments have shown that we remember pleasant experiences of childhood more than unpleasant ones. This is why it is essential that you do everything possible to make student learning of your material as pleasant as possible. You remember some unpleasant experiences because you enjoy telling them to your friends, not because they were unpleasant.

c. <u>Interference Learning</u>.

(1) The psychologists call this retroactive inhibition. For example, you learn to drive a car with regular steering. Later, you learn to drive with power steering. When you step into the car with conventional steering again the learning of power steering has interfered with the old skill you had and you must relearn to drive with conventional steering.

(2) Interference learning has implications in studying for an examination. Don't study different subject material between the time you do your studying and the time you take the test. Psychologists have found that you forget less when you sleep through eight hours than when you are awake for the same period, since new impressions gathered while awake are interfering with the older material learned. Study for your exam before you go to bed and you will be best prepared to take it first thing the next morning. If you have an afternoon exam, study or review during the noon hour.

Chapter II

METHODS OF INSTRUCTION

A. GENERAL

1. PURPOSE. This chapter is designed to give the marksmanship instructor a working knowledge of the methods of instruction used, the advantages and limitations of these methods, how to conduct effective instruction using the several methods, and broad policies relative to methods of instruction.

2. INSTRUCTIONAL POLICY. Since all marksmanship instruction is aimed toward effective student performance, emphasis will be placed upon practical work rather than upon theoretical instruction. Lessons should be reviewed periodically to insure that they include a maximum of student performance activities. Conference techniques should be used to the maximum for all oral presentations. Instructors should lecture throughout the whole instructional period only when no other method of instruction is appropriate. In all instruction, the accent is on student learning, and the students learn best by participating actively in instruction and accepting their share of responsibility for the teaching-learning success. Because of the diversity of training, recognize that no one method will suffice to accomplish all types of training. The method must fit the student learning outcomes desired. If the objective is a skill, then you must include demonstration and ample student performance. If the objective is a general knowledge or understanding, then you could use a combination of lecture and conference. The various methods are classified below to assist explanation. There is no sharp line of separation among methods; usually the instructor uses a combination of methods to achieve the lesson objective.

3. PRINCIPAL METHODS OF INSTRUCTION.

   a. Lecture. The lecture is a method of instruction in which the instructor orally develops his subject without student discussion. Because of the lack of student participation, the lecture should be used for an entire period of instruction only when no other method or combination of methods is appropriate. Normally, instructors lecture during most periods of instruction to give directions, present lesson introductions and conclusions, and to guide or summarize student discussion.

   b. Conference. The instructional conference is a method of instruction by which the teaching points are developed primarily through student discussion. The instructor initates student discussion by asking thought-provoking questions. He guides discussion through timely use of follow-up questions, subsummaries and transitions. The emphasis during the conference is to achieve learning through student participation. The conference uses a combination of lecture, demonstration, and performance where appropriate to assist student understanding.

   c. Demonstration. The demonstration is a method of instruction which assists student learning by showing correct procedures and expected standards.

   d. Performance. The performance method emphasizes student practice and application of principles and procedures. Students learn much by doing. Learning is frequently not complete without practice. Student performance consists of problem exercises to solve or routine practice to correct errors and acquire skill.

   e. A Concept of Method. As Figure 2-1 illustrates, in the telling or lecture method, student participation is at its lowest. In the performance method, student participation or activity is at its highest.

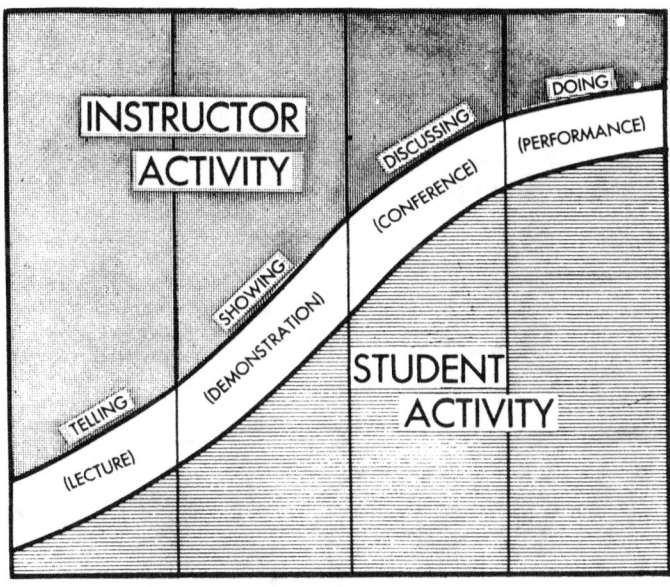

Figure 2-1. A Concept of Method.

B. LECTURE

1. THE "TELLING" METHOD. In the lecture method, you tell facts, principles, theories, or relationships which you want your student to learn and understand. When the word lecture is mentioned, students and instructors immediately think of a very formal discourse on a subject read by an expert from a prepared manuscript. Such a formal lecture is only one type of telling. They overlook that a lecture can be delivered by memorizing the key ideas with only occasional glancing at notes. It could even be extemporaneous or impromptu telling. The lecturer may have check-up questions planned and asked at appropriate intervals. He may encourage student questions during his telling. Essentially, the lecture is one-way oral presentation of ideas. Accordingly, the lecture uses an instructor-centered approach since the teller plans no (or limited) student participation. Some instructors regard any use of questions by the principal instructor as evidence of an instructional conference, but this is incorrect since the primary emphasis is still upon telling.

2. USE OF LECTURE. Many persons believe that lecture is an unproductive method of instruction. This is true only if the presentation is so dull that student attention wanders. As an instructor, you will lecture to introduce, subsummarize, and conclude your presentation, to direct and critique student performance, to narrate demonstrations, to explain and illustrate principles and procedures, to give examples, testimony, definitions, and statistics in support of teaching points. The lecture in combination with other methods of instruction is an indispensable ingredient of effective instruction.

3. ADVANTAGES OF THE LECTURE.

a. With the lecture you have an effective method of presenting many facts and ideas in a short period of training time. You are in control of the time and after rehearsals can make optimum use of this time. CAUTION: You must be careful not to "run ideas by" so fast that you confuse the student. You can minimize this danger by checking on student learning frequently, by providing opportunities for students to question points they do not understand, and by soliciting such questions when you have completed a complex teaching point. If student answers reveal a general lack of comprehension, you will have to rephrase your ideas, add substantiating material or further examples, and then recheck comprehension.

Figure 2-2. The Lecture: One-Way Presentation of Ideas.

    b. Place emphasis specifically where you desire it. Since you control the learning situation closely, develop only those points you need to develop. Through your experience and research into the instructional problem, you select the important facts to emphasize and subordinate less important facts and ideas.

    c. The lecture is very effective in presenting new or complex information to provide students with the background of knowledge they will need to participate in subsequent discussions or in performance exercises.

    d. The size of the listening audience does not restrict the lecture. You can deliver the lecture to twenty, two hundred, or two thousand students.

e. The lecture is an effective method for guest speakers or those who present constantly changing material. It is most effective when presented by a real authority or subject matter expert. When the speaker is well informed and uses good presentation techniques, he can hold the interest and stimulate the imagination of his listeners. The speaker who projects an interesting personality into his presentation, encourages student questions, periodically checks student interest, uses training aids properly, and arranges his material in an understandable way, will convey his message effectively.

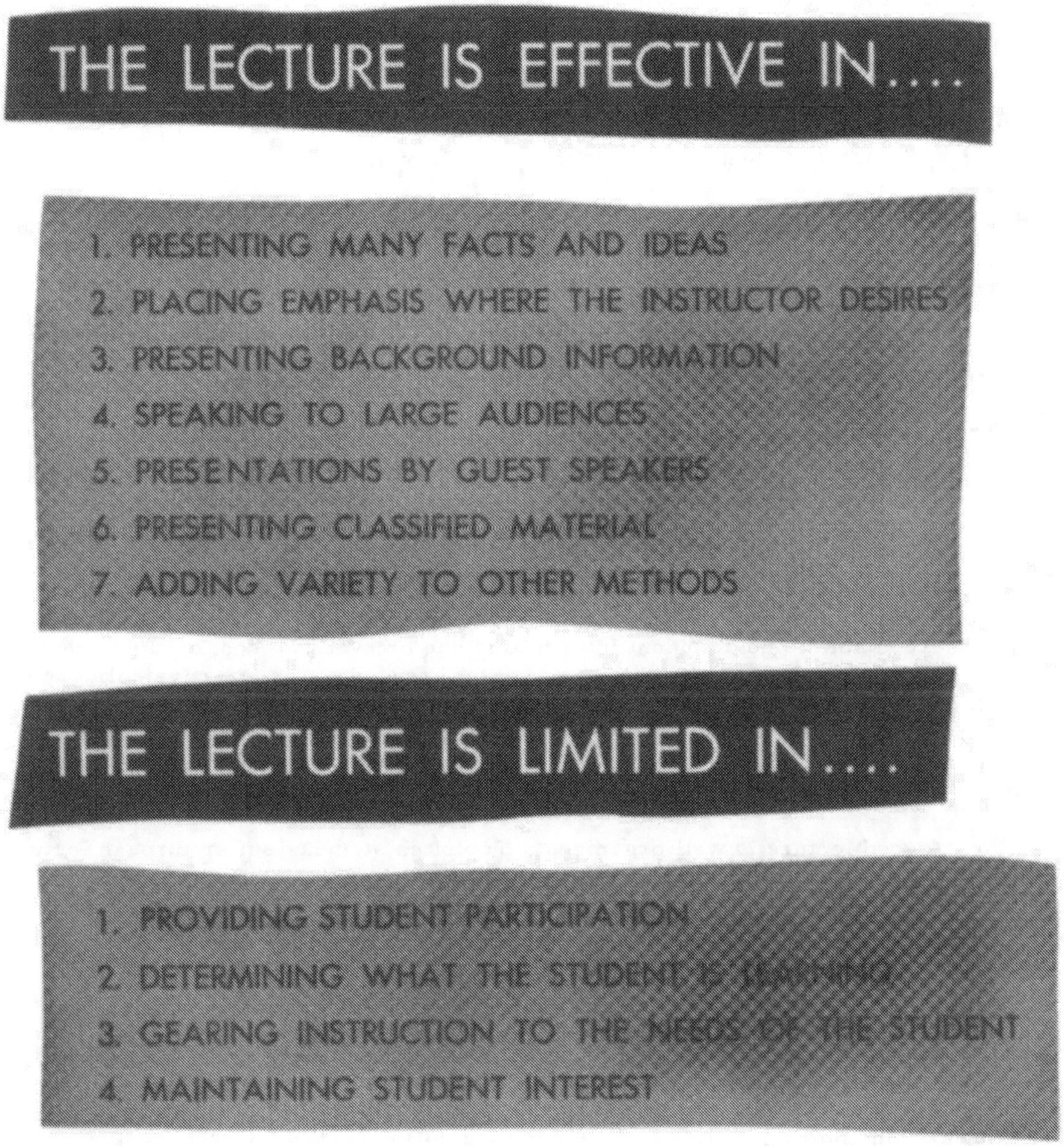

THE LECTURE IS EFFECTIVE IN....

1. PRESENTING MANY FACTS AND IDEAS
2. PLACING EMPHASIS WHERE THE INSTRUCTOR DESIRES
3. PRESENTING BACKGROUND INFORMATION
4. SPEAKING TO LARGE AUDIENCES
5. PRESENTATIONS BY GUEST SPEAKERS
6. PRESENTING CLASSIFIED MATERIAL
7. ADDING VARIETY TO OTHER METHODS

THE LECTURE IS LIMITED IN....

1. PROVIDING STUDENT PARTICIPATION
2. DETERMINING WHAT THE STUDENT IS LEARNING
3. GEARING INSTRUCTION TO THE NEEDS OF THE STUDENT
4. MAINTAINING STUDENT INTEREST

Figure 2-3. Advantages and Limitations of the Lecture.

f. The lecture adds variety to other methods of instruction. A dyamic instructor can lecture certain details or teaching points in an extremely interesting way. This provides a welcome change of pace when interspersed among other methods. It will provide a "time cushion" since you can streamline your explanation or amplify it as training time permits.

4. LIMITATIONS OF THE LECTURE.

a. The major limitations of the lecture are the lack of active student participation, lack of complete free two way communication between instructor and student, and the problem of maintaining high student interest level. Lectures foster overdependence upon the teacher as the expert and encourage student passivity and uncritical acceptance of everything the speaker says.

b. It is relatively difficult to determine what the student is learning. In a straight lecture, you are not sure whether the student grasps the material or not. By using check-up questions, you only sample class comprehension. Those you called upon may be the outstanding students or the poorest students in the class; this is a matter of chance.

c. You have difficulty in gearing your instruction to the needs or progress of the student. Usually you plan your presentation one way and present it that way. You have little or no flexibility to adapt to the learning level of the students as you proceed.

d. Student interest is difficult to maintain. It requires a skillful, dynamic speaker, aware of the need for control of class interest, with a fine change of pace to maintain class interest for extended periods of time. One critic stated it: "The mind can absorb only as much as the seat can endure."

5. GOOD LECTURE TECHNIQUES.

a. In the lecture, students learn by what they hear. Therefore, it is extremely important that you present your ideas in an interesting, enthusiastic manner. Student interest is difficult to maintain in a lecture. For this reason, keep your lecture time brief by varying lecture with conference, demonstration, or student performance. You have no doubt listened to lecturers who were able to hold student interest for one or even two hours. Perhaps you attribute this ability to the interesting personality, vast knowledge, or effective delivery of the instructor. These qualities are of course important, but if you listen carefully, you will usually discover that this interesting lecturer plans for and liberally uses the interest factors discussed in Chapter 5. He tells stories, he may use humor, and his presentation is always sprinkled with "for example. . ."

b. A good lecture is well organized. Too often students suffer through lectures which are rambling presentations of a mass of apparently unrelated details. They deal in generalities and never seem to come "down to earth." As an instructor, you can use the lecture to present vast amounts of detail, but you have no assurance that the student will absorb the details, much less remember them. By organizing your lecture around a few teaching points or main ideas and logically relating the details to these main ideas, you can make it much easier for the student to learn and retain the information. After you have explained and illustrated each main idea or difficult point, summarize before you go on to the next point. During your subsummary, give students an opportunity to ask questions. Ask questions yourself to insure that they understand the point. In going to the next teaching point, make a good transition. Too often we hear "And now we will. . . ." A good transition not only alerts the student to the fact that we are leaving one point and beginning another, but it also shows the student the relationship between points and maintains his interest at a high level. A teaching vehicle will help to make transitions both interesting and effective.

  c. Since the lecture does not provide for active student participation, you must stimulate the student's sight as well as his hearing. Use training aids which not only reinforce learning, but help to maintain student interest in your presentation.

  d. Another excellent technique is to provide a skeleton outline handout with ample room for students to make notes as you present. Pass this out before you begin. Students will learn better if they are actively engaged in following your presentation.

  e. Usually you can improve your lecture by asking two or three thought-provoking questions; these enable you to introduce some discussion. Many marksmanship subjects lend themselves to an interesting demonstration or problem-solving exercise. Combine the lecture with other methods to increase student participation and overcome this major disadvantage of the lecture method.

 C. <u>CONFERENCE</u>

  1. <u>THE "DISCUSSING" METHOD.</u>

   a. As education has placed increased emphasis on student activity for optimum student learning, the instructional conference has proved effective in securing good student participation. As a leader of an instructional conference, you will discover that the conference is not an easy way out. You must research every facet of your subject because new questions will arise from students in each succeeding discussion. Your teaching task will require that you exercise a high degree of intelligence, tact, alertness, and ability to think on your feet. A poor lecture is at worst boring; but a poorly led conference can be devastating when measured in terms of mislearning, student antagonism or frustration.

   b. The word "conference" has many meanings. In industry or in the military, it is "an assembly of a group of individuals in which the group members contribute information and ideas toward accomplishing a common purpose." The goal or common purpose of the conference may be a problem to solve, a decision to make, or learning or information to be shared. Regardless of the goal, the outcome of this type of conference is not preconceived or preplanned by the conference leader. Each member contributes in theory to the group discussion, solutions, or conclusions. In short, the conference leader does not attempt through skillful manipulation or guidance of the discussion to lead the group towards preconceived solutions or learning.

   c. In marksmanship training, we must standardize doctrine and administrative procedures. Since this is the case, how can we possibly use the conference method to teach students doctrine and procedures? Not only is the objective of each lesson predetermined but so also is each teaching point. The key to the answer is that the teaching points are developed primarily by student discussion. The instructor initiates, stimulates, and guides this student discussion. When sufficient pertinent discussion occurs, certain conclusions can be drawn from this discussion. These conclusions are the principles, teaching points, or doctrine. The method is actually inductive: from a series of student opinions, solutions, or experiences, the class builds up the principle to be learned. In the lecture method, the instructor frequently uses the deductive process; he states a principle or teaching point and then proceeds to prove it by a series of examples, illustrations, statistics, or other supporting material. The technique in conference leadership is to stimulate the free flow of ideas between instructor and students. Discussion is the catalyst for student understanding.

   d. To understand the conference method as it is used, consider this question: What is the difference between the instructional conference and the recitation method? In the recitation method, the instructor asks only check-up questions which require verbatim or memorized answers which re-echo the study materials. In the instructional conference, the instructor stimulates thought and discussion with questions designed to bring out different opinions, experiences, and techniques so they may be analyzed, compared, and discussed. A good instructional conference has stimulating lead-off, follow-up, and check-up questions planned for use at appropriate times during the lesson.

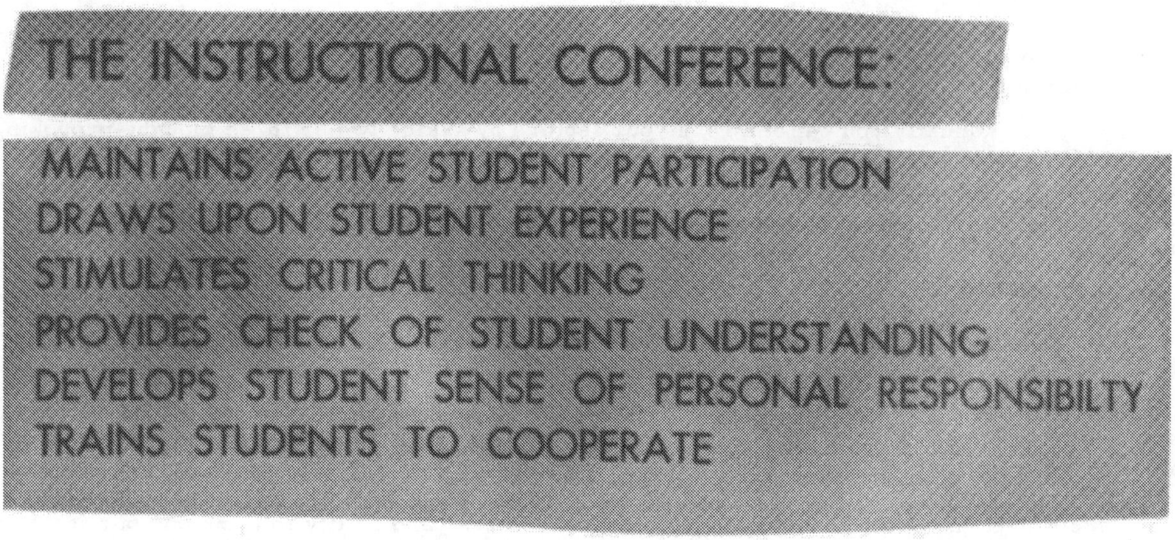

THE INSTRUCTIONAL CONFERENCE:

MAINTAINS ACTIVE STUDENT PARTICIPATION
DRAWS UPON STUDENT EXPERIENCE
STIMULATES CRITICAL THINKING
PROVIDES CHECK OF STUDENT UNDERSTANDING
DEVELOPS STUDENT SENSE OF PERSONAL RESPONSIBILTY
TRAINS STUDENTS TO COOPERATE

Figure 2-4. Advantages of the Instructional Conference.

2. ADVANTAGES OF THE CONFERENCE.

    a. The conference or guided discussion method encourages active student participation and maintains interest. When students discuss, probe, disagree, or answer provocative questions they are concentrating, thinking, and learning actively. Such active learning makes a greater mental impression and is remembered longer.

    b. During the guided discussion students often contribute new ideas and new applications from their background of experience. No instructor no matter how well informed can match the cumulative wealth of experience of the class. These student-originated ideas not only make instruction more meaningful, but they result in course improvements.

    c. Discussion stimulates reflective thinking and reasoning. Students become accustomed to thinking critically, to making comparisons, and to relating ideas and doctrine to their experiences and previous learning. The result of a good conference is that students consider all aspects of the problem. Many points which might otherwise still be doubtful are resolved by this questioning and discussion.

    d. The conference method provides you with frequent opportunities to check student comprehension of the subject. This enables you to gear instruction to the proper learning rate for that group. You can get an immediate reaction to how well students are absorbing the materials to be learned.

    e. Since active student participation is essential for an instructional conference students must assume responsibility to assist in their own learning. The burden of learning responsibility shifts from the instructor's back and becomes a shared cooperative task. Both the students and the instructor must insure that learning takes place.

f. A bonus effect in the group participation is that the conference trains students in the skills of cooperative effort, group thinking on a common problem, self-expression, and tolerance of the opinions of others.

3. LIMITATIONS OF THE INSTRUCTIONAL CONFERENCE.

a. The instructional conference method consumes more training time than does the lecture to cover a specified amount of instructional material. Ideas must be discussed, analyzed, accepted, modified, or rejected. CAUTION: Make very sure that you have selected for conference only those portions of your lesson that are extremely significant, controversial, or difficult to learn or accept. Don't waste time on a conference on the obvious or on material which does not permit several points of view. For example, beginning instructors sometimes try to conduct a conference on the organization of a team. This is a fact and no discussion is possible. However, a discussion might be conducted on whether this organization is of maximum efficiency.

b. The conference method is employed for classes containing up to one hundred students with the belief that it is better to have some discussion rather than none. It is felt that even limited student participation will bring out the major differences of thought or usual difficulties. It is most effective in small groups of fifteen or twenty students where all students voice their opinions. When you use the conference method with one hundred students you only sample student opinion and only the more aggressive students will volunteer to express their opinions.

c. To conduct a guided discussion you must have students with some knowledge of the subject to be discussed. Your advance study assignment should help to furnish their basic background knowledge. Advance sheets must provide the review of principles, help to get the student "read into" the problem.

d. Many gaps in learning or even some incorrect learning may occur if the instructor is inexperienced in planning and conducting a good conference. Compared with the lecturer, the conference instructor must possess more comprehensive knowledge of the subject more tact, greater versatility and flexibility, and an ability to think on his feet to seize the key points of student discussion and to summarize skillfully.

4. PLANNING A CONFERENCE.

a. Planning the introduction and conclusion for an instructional conference is similar to planning of any lesson and is discussed in Chapter 3. The primary concern in this paragraph is planning for discussion in the body of the lesson.

b. Since a discussion presupposes an adequate student background knowledge of the subject matter, you must analyze student background in terms of previous instruction and previous experience. For example, you know that few novice shooters have previous experience. You can, however, take advantage of the instruction that these students have received to date, as well as their backgrounds. You can supplement the student's knowledge by issuing advance study assignments, by lecturing initially, or by using both techniques.

c. In preparing your lesson plan, consider the teaching points as the framework of the conference, with each teaching point constituting a potential discussion area. Prepare questions to provoke thought, to stimulate discussion, and, above all, to insure thorough understanding of the teaching points. Avoid the tendency to ask only "what" questions. Have the students consider also "why", "how", "when", or "where". For example consider the discussion potentials of these questions: "What is the primary function of the gas system?"

(NOTE: For a treatment of types of question and their characteristics, see Chapter 6)

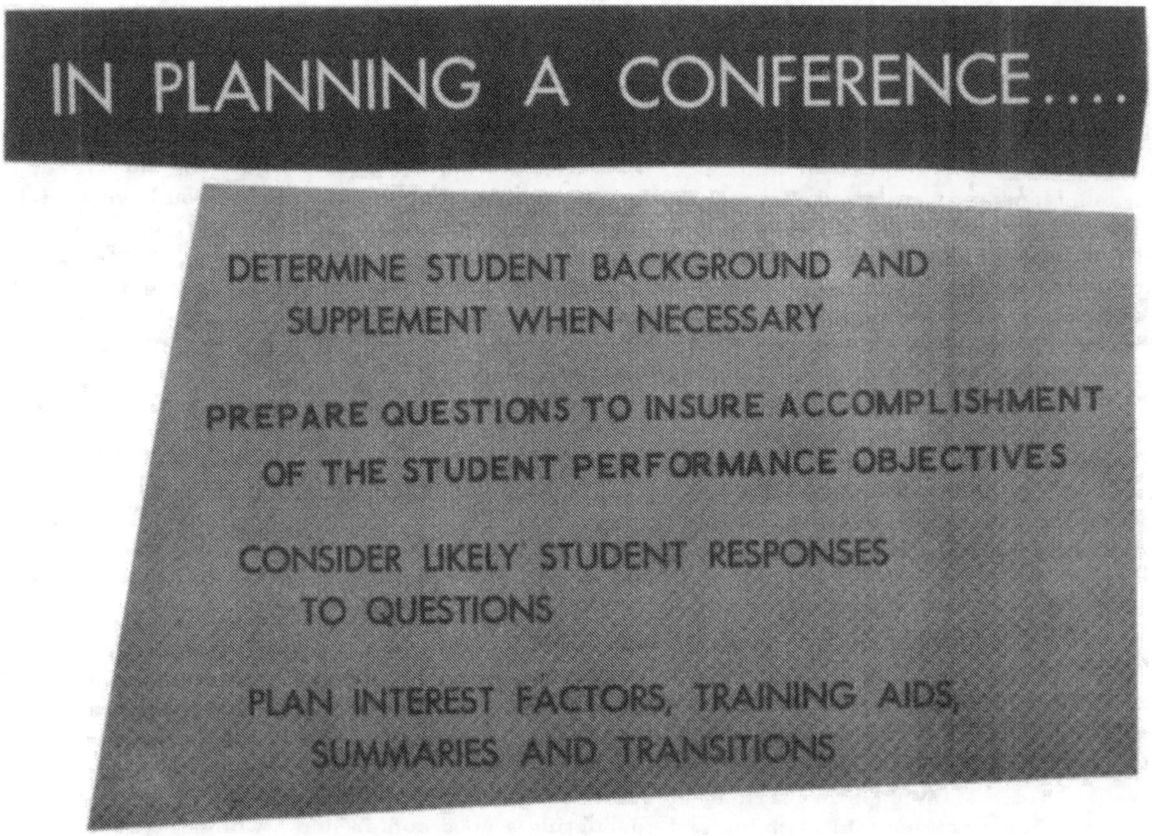

IN PLANNING A CONFERENCE....

- DETERMINE STUDENT BACKGROUND AND SUPPLEMENT WHEN NECESSARY
- PREPARE QUESTIONS TO INSURE ACCOMPLISHMENT OF THE STUDENT PERFORMANCE OBJECTIVES
- CONSIDER LIKELY STUDENT RESPONSES TO QUESTIONS
- PLAN INTEREST FACTORS, TRAINING AIDS, SUMMARIES AND TRANSITIONS

Figure 2-5. Planning a Conference.

d. Taking each of your prepared questions in turn, consider likely student responses to these questions. While student responses vary with the individual, they will usually fall into general response patterns. By considering likely student responses, you are better able to ask good follow-up questions and to lead the students to a thorough understanding of the teaching points. It is impossible for you to anticipate every student response; therefore, you must display versatility and ability to think on your feet when conducting a conference. You must evaluate student responses as they occur and use them to maximum advantage in advancing the conference or insuring student learning.

e. As with other instructional methods, a good conference requires a skillful use of interest factors, training aids, summaries, and transitions. Although questions are good interest factors in themselves, you will frequently find a need for examples, illustrations, and training aids to stimulate discussion and to support eaching points. Consider whether a teaching vehicle might assist student understanding. (See Chapter 5.) Frequent subsummaries are especially necessary in a conference. In the planning stage, you will not be able to determine the frequency with which subsummaries will be needed nor the comments that will occur during discussion; however, you can plan for summaries of teaching points and important supporting ideas, for you know if the discussion has been long and animated there must be a subsummary before moving to a consideration of the next teaching point. Similarly, you can plan transitions between important ideas, and deliver other transitions extemporaneously.

f. Plan advance sheets which contain thought-provoking questions or the initial discussion of the general situation so that valuable class time is not spent in preparatory reading.

5. CONDUCTING A CONFERENCE.

a. *Student Participation.* As a student you have probably attended an instructional period listed as a "conference" wherein the instructor could not get a discussion started. At no time did a student volunteer an answer, nor did students appear interested. The instructor was not able to secure student participation for maximum learning. This instructor probably attributed the lack of student particpation to a "bad day" or to a "slow class." It is more likely that the instructor either failed to understand or lacked the ability to apply effective conference techniques. Proper use of the techniques which follow will help you to arouse student interest and to obtain optimum student participation and learning.

b. *Permissive Atmosphere.*

(1) A conference is based upon student participation, and is most effective when all students are motivated to think reflectively and to enter into the discussion. Students will think more freely and will enter into a discussion more readily when you establish a permissive atmosphere, wherein you encourage a free flow of ideas. We do not have a single word in our language which aptly describes this relationship between instructor and student, so we go to French and we refer to instructor-student rapport: a feeling of mutual cooperation and understanding, of harmony and congeniality. You establish rapport by your sincerity and enthusiasm in your introduction, in your conduct of the discussion, and even in your administrative directions. You maintain rapport by the manner in which you ask questions, call upon students for contributions, and give credit for student ideas.

(2) For example, if you ask a question, pause, and then bark out a student's name in a drill sergeant's command voice, the student is likely to jump up, give the verbatim answer that he thinks you are looking for, and sit down. He has merely attempted to recall an answer from memory and has not attempted to think reflectively and come to a logical conclusion. This has the earmark of a grade-school type of recitation, not of a discussion among mature individuals.

(3) You can often stifle a good discussion by careless unthinking remarks. For example, you ask a question and call upon a student, who gives you a good answer, but one that you had not planned for. If, as happens all too often, you reply, "That is true, but it isn't what I have in mind," you are stifling student thought. You are, in effect, telling the student not to think, but to give you the correct response, to guess the "right" answer, to read your mind. In evaluating student responses keep in mind your objective - student learning - and make every attempt to promote such learning by stimulating student thinking and free expression of ideas.

c. *Stimulating Discussion.* By planning good thought-provoking questions and by establishing rapport with students, you have laid the foundation for discussion. You will find, however, that discussion will falter unless you take definite steps to stimulate further discussion. By making a startling or controversial statement or quotation, followed by a good question, you can often obtain enthusiastic participation once the students are aroused by these devices. Student stimulation by good questions is a technique used so successfully by the ancient Greek philosopher Socrates in his teaching.

d. *Guiding a Conference.*

(1) Based upon an analysis of the lesson objective and the teaching points, you have planned thought-provoking questions, questions to check student learning, subsummaries, and transitions. Within this framework, you must guide the discussion so as to insure student understanding of each teaching point and thus ultimately attain the lesson objective. How you guide and control the discussion is largely extemporaneous. Based upon student answers or contributions, you will frequently ask questions requiring further explanation or elaboration

to insure student understanding. Use this technique when a student is obviously bluffing or when a student uses terms which you realize are unfamiliar to some of the other students. Don't generally permit a student to make a bold statement without drawing from him his reasons for believing that way. In this way both you and the class can evaluate the soundness of the belief.

    (2) After you are successful in stimulating a good discussion, your next role is that of moderator. You must keep the discussion on the teaching point. When the discussion is obviously beginning to go "off on a tangent," summarize the comments and, by asking a question, bring the discussion back to the subject. Insure that participation is distributed among the students, that a few of the more enthusiastic students do not dominate the discussion. By checking off each student contributing on a class roster, you can single out nonparticipants and bring them into the discussion. When two students threaten to monopolize a discussion by getting into an argument, stop the argument by summarizing the opposing points of view, and bring a third student into the discussion for his opinion. Remember that at the end of each discussion your students expect you to voice the instructors position on this matter or summarize what they as students should take away as a result of the discussion.

    (3) Your contribution to student understanding depends largely upon the effectiveness of your subsummaries. A good subsummary of the students' discussion will briefly point out the ideas expressed, resolve conflicting points of view, relate ideas to the teaching point, point out application of principles, and state the instructor position, as appropriate. While subsummaries are planned for, their frequency and their content will depend upon the length and scope of the discussion. Subsummaries give an instructor an excellent opportunity to make transitions into his next teaching point or supporting idea. (For further discussion of subsummaries and transitions, see Chapter 6.)

    (4) Your role as discussion moderator is not an easy one. You must be firm in keeping the discussion on the subject, yet tactful in your control. A mark of a good conference leader is his ability to guide discussion toward desired learning outcomes without dominating the discussion. He realizes when there is need for a definition of terms, clarification, arbitration or summary and sees that these needs are filled.

NOTE: See Chapter 6 for specific techniques used in asking questions, handling student response, accepting voluntary contributions, and handling student questions.

 D. THE DEMONSTRATION

   1. GENERAL. Teaching is a process of telling, showing, and doing. Demonstrating, or showing, is used in combination with other methods, and is usually preceded and accompanied by lecture and conference and followed by student performance, critique, and conference. The student must know what to observe or look for in the demonstration. When your lesson objective involves the development of skills or the practice of procedures, your demonstration will be followed by student performance.

   2. PURPOSES. While the demonstration is usually regarded as a forerunner of student performance, it can also be used very effectively in supporting teaching points. Sometimes a demonstration is used purely to illustrate or support a teaching point. Many of the world famous Infantry School demonstrations are for this purpose. For example, such demonstrations as the Infantry Tank Team in the Attack, Firepower, and Airmobile Assault serve to illustrate the capabilities and tactical use of Infantry weapons and equipment. They are not designed to be followed by student performance or application. Uses of the demonstration are as follows:

a. To teach manipulative operations (how to do it).

EXAMPLE: Use a demonstration to give students a visual impression of how to disassemble the .45 Cal pistol. Demonstrate step-by-step. The combination of seeing the demonstration and hearing your comments reinforces student learning through the use of two senses.

b. To teach principles and theories (why it works).

EXAMPLE: Fire a pistol suspended on cords to demonstrate the recoil principle.

c. To teach operation and functioning (how it works).

EXAMPLE: Use a cutaway weapon to demonstrate the phases in the cycle of operation.

d. To teach team organization during team matches.

EXAMPLE: Have demonstration shooters assume positions to illustrate their important features.

e. To teach procedures (how men work together).

EXAMPLE: Use a skit to portray range safety and procedures.

f. To teach appreciations.

EXAMPLE: Have demonstrators fire weapon for accuracy to develop an appreciation for the skills involved.

3. ADVANTAGES OF THE DEMONSTRATION.

a. Demonstrations save time. A brief demonstration of the proper method of applying trigger pressure is more effective than a lengthy discourse. Doing is usually simpler than telling. To illustrate this, try to explain briefly, with words alone, how to tie a shoelace or how to make a Windsor knot in your tie. Notice how much clearer and simpler it is to demonstrate. Demonstration will make your explanation more concrete in the minds of your students.

b. Demonstrations insure thorough understanding through their appeal to sight as well as to hearing. They clear up student confusion through their appeal to several senses.

c. Demonstrations stimulate interest and therefore learning through their realism dramatic appeal, and the variety which they add to other methods of instruction.

d. Demonstrations set the stage for student performance by illustrating correct methods, by setting standards of performance, and by giving the student confidence that attainment of the skill is possible.

4. LIMITATIONS OF THE DEMONSTRATION.

a. During the presentation period, demonstrations do not normally provide for active student participation. You can overcome this limitation by asking questions of students and by encouraging student questions between steps in a demonstration, and by providing for student performance following the demonstration.

b. The more elaborate demonstrations require additional personnel and equipment and time for rehearsals. The increase in student learning, however, will more than compensate for the expense involved since demonstrations can be presented to large groups. Some Infantry School demonstrations are presented to a thousand students simultaneously.

c. Range demonstrations are frequently affected by weather conditions. Indoor substitutes are usually inferior or less effective.

5. FORMS OF THE DEMONSTRATION. There are five general forms of the demonstration:

a. The Procedural Demonstration. This is the form of demonstration used to show and explain the operation and functioning of weapons. This type of demonstration may be conducted indoors or outdoors and is used widely throughout basic and advanced training.

b. Displays. Displays must be planned so that students can view them quickly. This requires arranging the displayed materials so that each item can be seen by all students at the same time. For large classes use duplicate displays or divide the class into sections, with the sections rotating from one exhibit to another. This is known as the "county fair" system.

c. Range Demonstrations. Complicated demonstrations can be shown one part at a time; later, the complete performance can be shown. One phase must be properly assimilated before the next phase claims the student's attention.

d. Motion Pictures. Training films provide ready made demonstrations by experts. Here the student has the opportunity to see internal workings of weapons, or matches being conducted - things he would otherwise have to imagine.

e. Skits. Instructors or assistants may act out operations or procedures. This form of demonstration has proved an effective means of portraying range safety. Skits guide student appreciations and attitudes. Skits may be designed to show the wrong way; however, you must insure that the right way is obvious, or show the correct way later. Skits must be carefully planned and smoothly presented; this requires repeated rehearsals.

6. PLANNING A DEMONSTRATION. The success of a good demonstration depends ninety percent upon planning and rehearsal, and ten percent upon execution. (See Figure 2-6.) Here is a checklist of things to consider.

a. Based upon the lesson objective and the specific teaching point, decide what to demonstrate. Limit the scope so that the demonstration has a specific purpose. A lengthy, involved demonstration will confuse your students. Remember, the essence of effective demonstration is precision, timing and conciseness.

b. If your demonstration involves several operations, list these operations and demonstrate them one at a time. If you wish students to learn more than one way of performing an operation, plan a separate and distinct demonstration for each method.

c. Prepare a scenario. Include an introduction, an explanation and a summary, and incorporate training aids (charts and mock-ups) to aid in explaining the steps being demonstrated. Attach the scenario as an inclosure to the problem lesson plan.

d. Make a list of the personnel and equipment needed for the demonstration and arrange for their availability for rehearsals and performance.

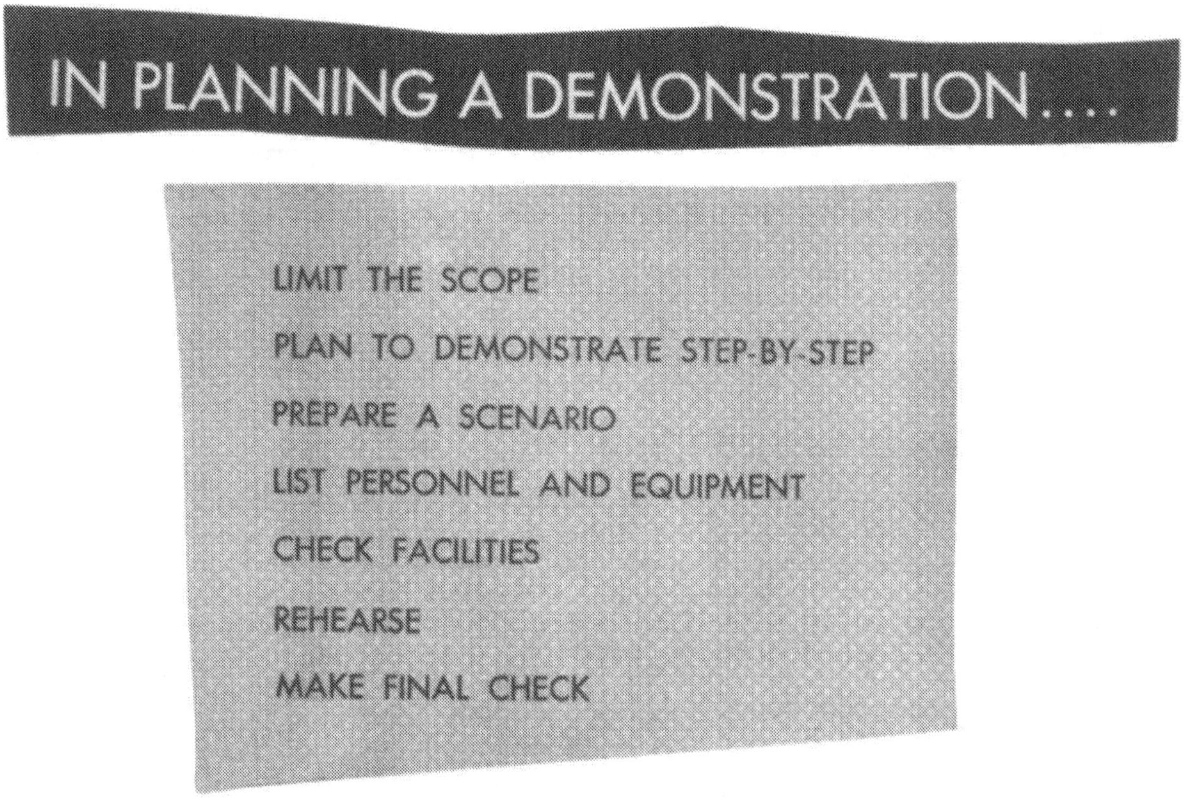

IN PLANNING A DEMONSTRATION....

- LIMIT THE SCOPE
- PLAN TO DEMONSTRATE STEP-BY-STEP
- PREPARE A SCENARIO
- LIST PERSONNEL AND EQUIPMENT
- CHECK FACILITIES
- REHEARSE
- MAKE FINAL CHECK

Figure 2-6. Planning a Demonstration.

 e. Arrange for and check the site of the demonstration. Insure that students can see and hear, and arrange for sound equipment if necessary.

 f. Rehearse the demonstration at the actual site to be used. Utilize initial rehearsals to check equipment and timing and to develop the proficiency of participants. Request the presence of other members of your team at your dress rehearsal to obtain constructive criticism and to insure high standards of performance.

 g. Prior to the actual performance, check to insure that personnel and equipment are present and that classroom arrangements are complete.

7. CONDUCTING A DEMONSTRATION.

 a. During a demonstration, you obviously want the student to watch the demonstration, yet you do not entirely forfeit your job of oral communication. If you are demonstrating an item of equipment, speak to the students, not the equipment. If, on the other hand, you are explaining while an assistant instructor demonstrates the equipment, direct the students' attention to the demonstrator. It is not uncommon for students to watch an instructor and fail to realize that an assistant instructor is demonstrating the steps of the operation.

 b. Insure that all students can see the demonstration. You may find it necessary to repeat a demonstration in a different position or at a different angle. Be careful that neither you nor your assistants block the students' view.

c. Introduce the overall demonstration carefully to insure that the students understand the purpose and what to look for. Knowing what they should get out of the demonstration affects the way students observe it and assists learning.

d. Explain the demonstration one step at a time. Introduce each step by telling the students what you will do next. During the step, explain what is being done and why it is done that way. Summarize after a complex step or after several steps during a lengthy demonstration. Ask questions to check student understanding. Encourage students to ask questions between steps, but do not allow students to interrupt the demonstration of a step.

e. Where sequence is important or where an operation involves several difficult steps, outline each step on the blackboard or with a chart before you explain and demonstrate the step.

E. <u>PERFORMANCE</u>

1. <u>STUDENT PARTICIPATION</u>. The performance method emphasizes student doing in order to learn and improve physical and mental skills. Sometimes it is routine practice of physical skills such as those used in weapons firing. Sometimes it involves problem solving skills in applying principles, techniques, or procedures to a realistic training situation. The performance method takes any one of two basic forms; the practice exercise and the problem exercise. You can introduce variety by combining these basic forms.

a. <u>Practice Exercise</u>. An applicatory exercise designed to teach procedures and to develop basic skills. The practice exercise is used widely to teach subjects such as maintenance and operation of weapons, and marksmanship. Variations of the practice exercise include:

(1) <u>Controlled Practice.</u> Practice by group in which group members practice the same thing at the same rate and time to gain the correct concept and to develop accuracy. The PI explains each step as the students perform. AI's control the practice and give aid. Controlled practice is especially suited for the initial step in learning a skill.

(2) <u>Individual Practice</u>. Routine practice by individual students until a skill becomes automatic. This practice is supervised, but not closely controlled since each student works at his own speed.

(3) <u>Coach and Pupil</u>. Used for teaching students who have learned the fundamentals of a skill or procedure. Both coach and pupil are students whose roles are periodically reversed. The coach and pupil technique is especially useful in marksmanship.

(4) <u>Team Practice</u>. An exercise in which students serve as members of a team. Team practice is normally conducted in two phases; first, a walk-through by the numbers in which techniques are emphasized; second, a phase allowing the application of techniques in a realistic situation.

b. <u>Problem Exercise: Characteristics</u>. The problem exercise employs the problem solving process and is used when solutions are based solely upon the application of principles, techniques, or procedures to the problem situation. Successive situations in a problem exercise may vary from a few minutes to several hours. It may be a simple situation presented orally and followed by questions, or it may be a written exercise involving considerable computation. The problem exercise can be employed in applying data or in teaching any subject in which computations are required.

2. CONDUCTING STUDENT PERFORMANCE.

    a. <u>Motivate</u>. Because a student is performing an act does not mean that he will learn well. Tell students specifically what they are to learn and why. Men will meet a challenge, so let them know what standards you expect of them, and tell them how well they are progressing. Maintain interest through competition. Overcome monotony by varying your procedures and by gradually increasing performance standards. Make performance realistic.

    b. <u>Explain and Demonstrate</u>. In explaining a procedure, use conference techniques and encourage student questions, not only to insure understanding but to make the student interested in learning the procedure. A good demonstration will make the procedure clear and will set standards for the student to attain.

    c. <u>Control and Supervise</u>. Make your directions specific, and encourage students to ask questions if they do not understand. Brief your assistants thoroughly so that they help you control student activity. Supervise constantly to insure that students understand how to perform correctly and are making satisfactory progress. When students learn to perform the operation correctly, insure that they are kept busy learning to perform it better and faster. Prevent students from forming faulty habits; but, at the same time, allow students to use initiative and resourcefulness. Your students will learn better if you and your assistants demonstrate that you are there to help them to learn not to harass them. For example, most men like to fire weapons, but too many are harassed into hating range work by oversupervision and over control. Keep your students wanting to learn and keep them challenged to do better.

    d. <u>Critique</u>. It takes common sense and tact to critique student performance properly. Frequently, both the student and the instructor are tired and exasperated by poor performance. You must encourage the student and help him to maintain self-confidence under the trying conditions of learning a new skill. At the same time you realize that a certain minimum standard must be attained to insure his safety. Skill in human relations is essential in this highly personal learning situation. Help, not harassment, is what the student needs. Praise the correct portions of performance. Select the major weaknesses to be improved first and offer definite suggestions how they can be improved. Let the student know periodically how he is progressing. As you detect definite steady improvement, bear down more heavily to achieve desired standards.

F. <u>VARIATIONS AND COMBINATIONS</u>

    1. <u>GENERAL.</u> In any training effort, methods have a way of becoming sterotyped. This is not wholly bad, in that the pattern of instruction becomes familiar to the student and requires less time in orientation and adjustment. On the other hand, variety of learning experience adds interest, challenges the student, and stimulates thinking. Variety for variety's sake is not the answer, however. The really effective instructor is one who possesses a thorough understanding of methods, techniques, and aids to instruction and who employs these tools of his trade to maximum benefit in each problem that he teaches. Ask yourself, "What is the best possible way that I can present this lesson?" and "In what setting can students learn this subject best?" The techniques described below can not only add variety to instruction, but they can, in combination with the basic methods of instruction, increase student learning. Some of these techniques are essentially combinations of the lecture and the conference, while others are characterized by special physical arrangement, timing, and the number of speakers A basic part of the USAIS training philosophy is the belief that training is so varied that it requires various methods of instruction to teach most effectively.

IN CONDUCTING STUDENT PERFORMANCE....

MOTIVATE

- STATE WHAT AND WHY
- SET STANDARDS
- EVALUATE PROGRESS
- PROVIDE COMPETITION
- VARY PROCEDURES
- INCREASE STANDARDS
- PROVIDE REALISM

EXPLAIN AND DEMONSTRATE

CONTROL AND SUPERVISE

- GIVE SPECIFIC DIRECTIONS
- HAVE ADEQUATE ASSISTANTS
- SUPERVISE CONSTANTLY
- KEEP STUDENTS CHALLENGED
- PREVENT FAULTY HABITS
- HELP - DO NOT HARASS

CRITIQUE AND SUGGEST

Figure 2-7. Conducting Student Performance

2. SMALL GROUP INSTRUCTION.

    a. Conferences and practical exercises are most effective when presented to small groups of students. Few instructors will dispute this statement, yet small group instruction is resisted on the basis that it requires more instructors (or more work for present instructors) in addition to a revision of existing facilities.

    b. For a fifty-man group, the traditional seating arrangement is modified by seating students at small tables arranged in four clusters, one group of twelve to fourteen students in each quarter of the classroom. Groups can be quickly separated by drawing curtains between them. When the instructor must have the attention of the entire class, the curtains can be drawn back, with students remaining in the same basic pattern. The use of tables facilitates any other arrangement which you might desire. In employing this arrangement, you can directly supervise the small group work, or, more commonly, as the principal instructor you may exercise overall supervision over student group leaders. The instructor provides student discussion leaders with background material and a discussion outline, in addition to the advance study assignments, which all students receive. Student recorders are selected to write the conclusions and decisions of the groups.

3. STUDENT COMMITTEE (WORKSHOP).

    a. Sometimes referred to as the workshop approach, study, research and student committees has proved very effective in graduate work at a number of civilian universities. The technique consists of breaking out a block of instruction into subject areas and assigning to student committees the responsibility for performing research in a specific subject area and reporting the results to the entire class. Normally, the student chairman will assign specific research tasks to each member of his committee and meet with his committee to resolve conflicts, arrive at conclusions, and prepare the committee report for presentation. The method of presenting the committee report can vary. For example, the committee chairman may present the entire report, then accept questions from the class and lead the discussion.

    b. A more effective procedure for increased participation is to have the chairman introduce the subject, followed by individual presentations by committee members, with brief conferences following each presentation. During the presentations, the instructor serves as a resource person. The committee method places responsibility on the student, who learns by his own research and activity. Since students will more readily modify their existing opinions when allowed to come to conclusions for themselves, the committee method lends itself to teaching new doctrine.

    c. Student presentations in the committee system are usually interesting and generate worthwhile conferences. The effectiveness of the method depends not only upon the instructor's ability to motivate the students initially and to organize the work, but also upon the sense of responsibility of the student chairman and that of the individual committee members. To insure worthwhile effort, civilian universities grade students upon their research papers, upon their presentations, or upon both. The method is not practical for large classes of more than fifty students. The method also requires many hours of student work for each presentation. The objection that only the student committee gets maximum benefit in each subject area can be partially overcome by requiring all students to sutdy advance assignments prior to committee presentations.

4. FORUM.

    a. Public discussions usually include forum periods, wherein audience members have a chance to participate. They can voice their opinions, question the speakers, or ask for elaboration on specific points. A lecture forum, for example, consists of a speech by a person with special qualifications, followed by a forum period. The lecture forum has application in club instruction following presentations by guest speakers, whose special knowledge and prestige will usually stimulate questions from the audience.

    b. You should not use the lecture forum in normal instruction, since students will seldom ask questions following a lecture. To illustrate what can happen, an instructor lectured to a class for 25 minutes. He then announced, "I have reserved ten minutes to answer the questions which I am sure you have." The ensuing silence was embarrassing to the instructor, who was forced to give his conclusion and dismiss the class early, which was exactly what the students had expected. The forum period is more successful following a symposium, debate, or panel discussion.

    5. SYMPOSIUM. The symposium consists of an introduction by the moderator (principal instructor), prepared presentations by two to four specially qualified speakers, (assistant instructors), a summary by the moderator, and usually a forum period in which the speakers answer questions from the audience. An example of a symposium is the popular "Town Meeting of the Air." Each speaker presents a particular phase of the topic or a different point of view. Each talk should be short and to the point (ten to twenty-five minutes). The symposium is formal in nature, and like the lecture, insures that much material can be covered in a short period of time. It may be used to present new material in a concise logical way. It permits several objective viewpoints of subject matter which may be controversial. Frequently, a symposium will clarify aspects of a knotty or difficult problem by giving the students a chance to consider each angle separately as each speaker presents the new viewpoint. If the speakers are concise they will keep the students stimulated to think and study and compare as the symposium progresses. Again, interest in the symposium is maintained by the variety of speakers who appeal to different students within the class. The symposium lends itself to any instruction in which student performance is not practical and where much factual material must be presented in a short period of time. A limitation is that student participation is at a minimum since there is only a limited time for student discussion. Sometimes questions are permitted after each talk or a question period is reserved at the end. In either case, however, the time is limited and a few questions might not always clarify the issues raised.

    6. TEAM TEACHING.

    a. Related to the symposium technique is the team teaching concept of "double lectern" approach. In this technique the instructional lecture is shared by two (or more) instructors. Each instructor is assigned a specific portion of the lesson to present. As the period progresses the instructors alternate as principal instructor with each instructor developing a complete teaching point or a specific block of supporting material.

    b. Such a system usually generates high student interest since it is a new or different approach. You must be careful not to jump from one instructor to the other too frequently or students will become confused. Although it has been said that "there are no dull subjects, just dull instructors" this is one way to enliven a subject of potentially low student interest.

    c. Modern civilian schools are employing this teacher task force or teacher work team concept successfully in the grade school and high school levels.

    7. DEBATE.

    a. An instructional debate is essentially a symposium limited to two points of view. Speakers should be approximately equal in experience and prestige. The participants consist of a moderator and two to four speakers. In the three-man debate, a speaker usually introduces the topic and presents the general issues, followed by two speakers who in turn present opposing points of view. In the traditional four-man debate, Speaker A presents his point of view, followed by Speaker B, who presents an opposing point of view. Speakers C and D in turn support the stands taken by A and B, respectively, but with additional reasoning. The moderator, as in a symposium, introduces the topic and speakers, controls the debate, summarizes the presentations, and opens the topic to discussion by the audience.

b. Because of its competitive nature, the debate has student appeal. The subsequent student participation may be limited, however, because of the feeling that the speakers, especially if they are known to be highly qualified, have left nothing else to be said. Debates between students, however, usually encourage active participation by other students.

c. A major disadvantage of the debate is its tendency to emphasize differences, making it difficult for the class to reach a consensus. A skilled moderator can overcome this handicap during the forum period. The principal instructor must insure that the student learning outcome is in terms of how to apply the different points of view to meet varying situations, rather than blind adherence to one point of view only.

8. CASE METHOD. The case method of instruction was popularized at the Harvard School of Business and is used extensively at personnel and management schools and seminars. As the name implies, students study and discuss principles of management, personnel, law, leadership, etc. through the medium of one or a series of cases. In each "case," the principal events and facts of the situation are outlined. Generally, the case involves a human relations situation - the action, reaction, or interaction of one person upon another person or group of persons. Sometimes students read this "case" as a homework assignment. The next day they are called upon to state what they believed was the cause for the interpersonal difficulty or what action they would now take as a result of the situation. Such action or opinion is then discussed by other members of the class and other proposals are advanced. There is no preferred solution, the whole purpose is to arouse an adequate discussion of principles involved.

9. PANEL.

a. A panel is less formal than the symposium. It consists of a chairman and two to four persons with special knowledge of a topic who converse in a logical fashion about the topic important to the class. The chairman guides the conversation by asking questions and by inviting members to contribute their ideas. The key to a successful panel is the apparently spontaneous flow of conversation between the members, with a minimum of control by the chairman. Panelists remain seated while speaking, make their contributions brief, refer to each other by name, and speak loudly enough for the students to hear. The conversation progresses according to an outline familiar to the members of the panel.

b. When members have made their contributions, the chairman summarizes and opens the discussion to student participation. Panels are most effectively employed to give a class an understanding of the several facets of a topic, to present advantages and disadvantages of a proposed solution to a problem, or to present varying points of view. Since the panel can bring the class up to date on a topic, it is better than the conference for introducing a new subject. Although panel members seldom have the opportunity to present their views completely, a well-planned panel can create interest in a topic as a prelude to a conference.

c. In using panel discussion, it is important that the chairman introduce the members by stating their special qualifications. A panel topic should lend itself to the presentation of varying points of view or to interrelated facets of the topic in order that a realistic conversation can take place.

10. SMALL GROUP DISCUSSIONS (SEMINAR).

    a. Small group discussions have one of these objectives: group decision, majority and minority decision, or individual decision or opinion. To accomplish these objectives, groups may be formally organized, partially organized, or without organization.

    b. In a formally organized group, each member is assigned a specific role. This type of group is used when the objective is a group decision. The student leader receives sufficient guidance to help him guide his group to a decision. He strongly guides the discussion to insure that the major facets of the problem have been considered. All students receive an advance handout outlining the major areas and points related to the problem. When, or if, the class reassembles, he, as group leader, is responsible for seeing that summaries of the group's decision are presented. He may summarize personally or assign the task to students who have had a major share in determining the decision and who believe strongly in it.

    c. Groups may be formed with varying degrees of organization. In such a partially organized group, the group leader may be designated to form a simple decision or several persons may be designated as the researchers for the group in special technical areas. Another version may be to designate a person to insure that a minority opinion is considered, discussed, and presented.

    d. In the group without organization, no roles are assigned. As the discussion develops, individual members begin to form individual opinions. As the need arises, members assume or commandeer certain roles or jobs within the group in order to accomplish the group task.

    (1) The objective may be a group decision by democratic majority rule, a majority-minority decision, or individual decision. Or, the objective may be to expose students to experiences in decision making by democratic process or to expose each individual in the group to as many facets of the problem as can be generated and considered. This will depend upon the experience level of the participants, how they interact with each other, and the training time available.

    (2) The student designated as discussion leader receives written guidance for successful group leadership. He never dictates the decision; he never gives the answers. He must insure that specific areas or points are considered and amply discussed, but he must do this indirectly; he must not teach these points.

    (3) If the objective is a group decision or solution, the rule of simple majority will determine the group position. The goal may be that each member arrive at his individual conclusion. If the objective is only an individual opinion he should see the problem area, conflicts, and factors which will influence a good decision. He has thought for himself. He had to express his ideas and lay them before the group for examination and debate. He arrived at a decision or opinion as the result of his own analysis or of the opinions of his peers. It was not school dictated or principal instructor dominated.

    e. Regardless of the type of group organization, the operating procedures are similar. Generally, the principal instructor briefly orients the whole class on the problem, the desired objective, and general procedures, then divides the class into small work groups. Most of the training time is spent in the work group. At the end of the group discussions the instructor usually reassembles the class to consider the areas discussed, the problems and conflicts which arose, and proposed paths of action. He may solicit group solutions, dissenting opinions or individual solutions, depending upon the general objective for the class and the time available.

f. All these types of small group discussions offer the advantage of a mature, realistic learning situation. Decisions are reached after careful analyses of factors involved in the problem. They are student centered with maximum student participation. They shift the burden of learning squarely upon the shoulders of the students. They train students to listen carefully, to work cooperatively, and to express opinions tactfully. They challenge student initiative. They stimulate thorough preparation, careful deliberation, and creative thought.

11. "DISCUSSION 66" OR BUZZ SESSIONS.

a. This technique, popularized by J. Donald Phillips of Michigan State College, is used in large educational conferences and business gatherings to allow each member of an audience to briefly state his opinion. The term "66" is derived from giving each six-man group six minutes to state opinions and come to conclusions. Of course, the actual figures can vary; for example, four-man groups could be given eight minutes. The chairman of each six-man group is given a card with a question or statement on which a group opinion is desired. Different questions are normally issued to the different groups. Each member of the group voices his opinion, which the chairman or a recorder places on the card. The chairman then determines from the group the opinion which they believe is the best conclusion. Sometimes both a majority and a minority opinion are reported.

b. The buzz session technique is frequently employed during the forum period following a presentation, in which each group spokesman is called upon to present his group's conclusion. It may be used, however, to formulate a conference agenda or to determine questions which members of the class hope will be answered by the speakers. The technique can be adapted to marksmanship instruction, especially in areas wherein there is no single preferred solution, such as a solution to some team coaching problem.

12. BRAINSTORMING. Another technique used in industry to exchange ideas is brainstorming, wherein participants are presented with a problem and are encouraged to contribute any solutions that come to mind, no matter how far fetched or impractical they appear to be at first glance. This very lack of emphasis on judicious thinking constitutes the real value of brainstorming. Participants are free to go beyond the mental barriers of conventional doctrine and contribute a variety of original ideas. The theory is that out of a hundred or more brainstorming ideas, further study may indicate that two or three ideas are worthy of adoption. Although brainstorming is more readily adapted to management than to instruction, it can be used to motivate students and to stimulate thinking in a subject area.

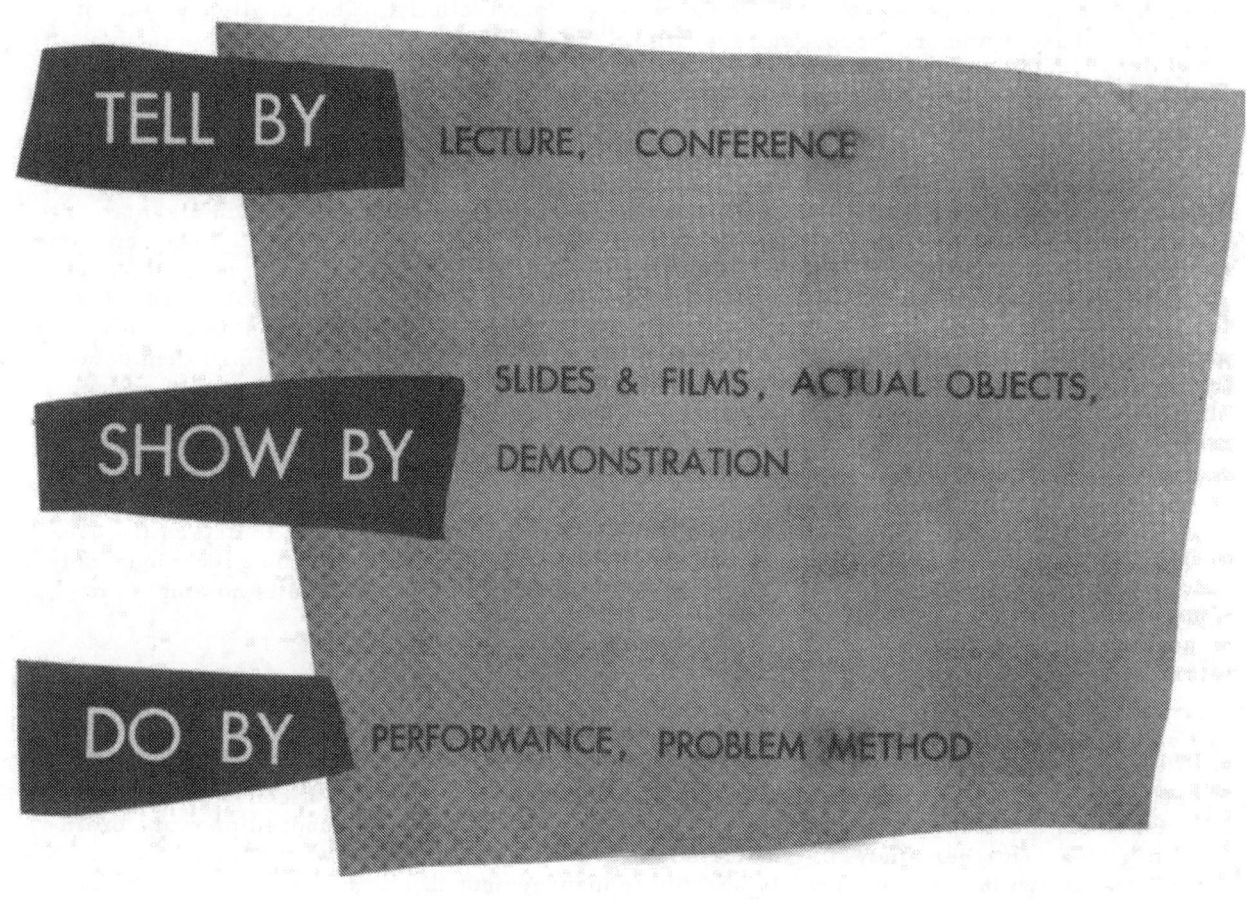

Figure 2-8. A Good Lesson Combines Several Methods

CHAPTER III

LESSON PLANNING

A. <u>GENERAL</u>: An outstanding and proper application of all material contained in this handbook is necessary to accomplish a high quality of instruction. Underlying the discussion of each instructional method, aid and technique is one basic premise--that the instructor has a complete, usable lesson plan. The heart of the usable lesson plan is the <u>lesson outline</u>; it is from this outline that the instructor prepares his presentation or lectern notes. The complete lesson plan includes the lesson outline, research materials, instructional handouts, copies or descriptions of training aids used, and the preparation data section listing physical facilities, personnel, sound, transportation, ammunition, and other requirements to present this lesson successfully. Lesson planning will occupy more time than any other single step in the instructional staircase (see Figure 3-1). Thorough understanding of lesson preparation is essential for all instructors.

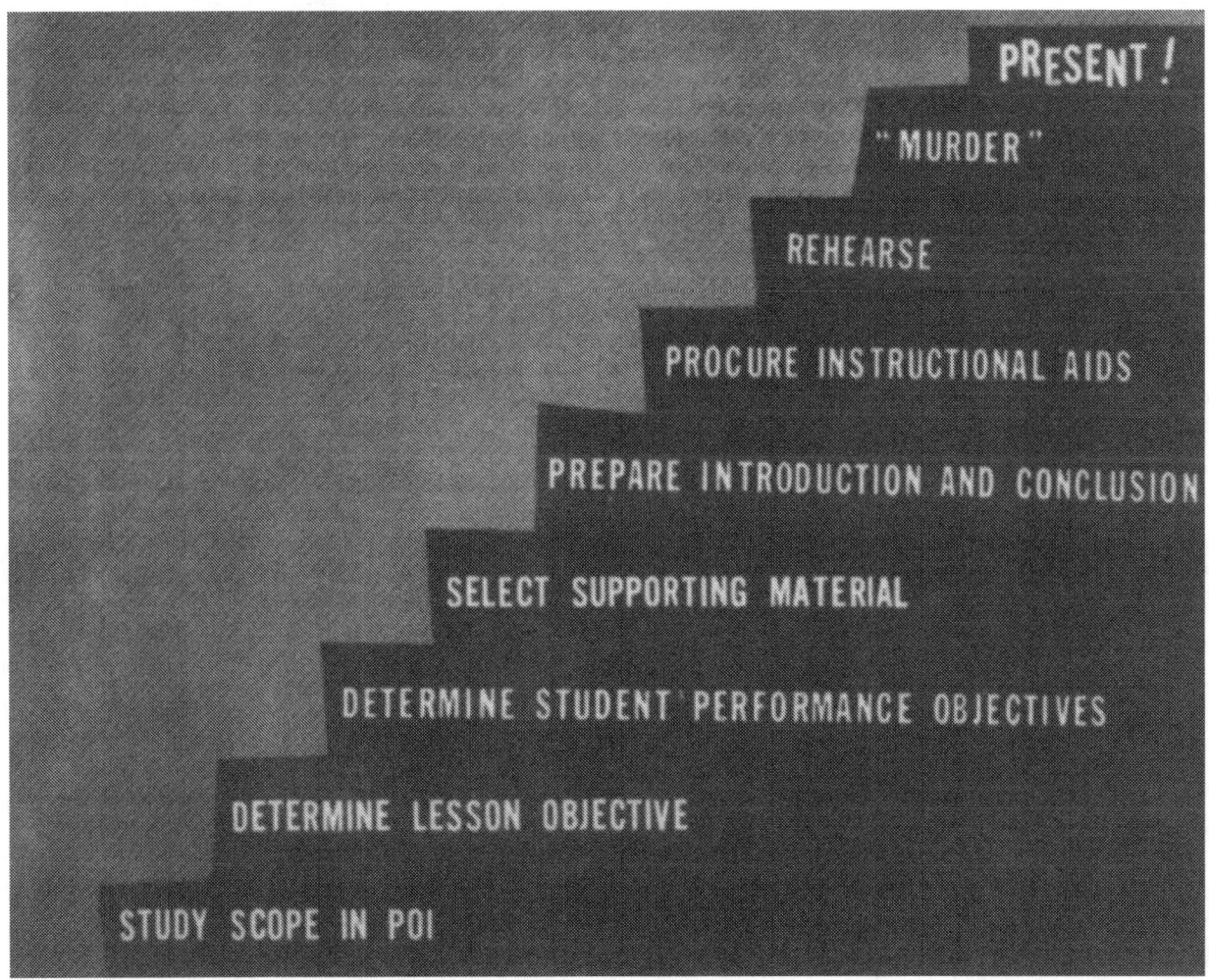

Figure 3-1. Instructional Staircase to Success.

1. ESTIMATE OF THE TEACHING SITUATION.

    a. <u>General</u>. In planning the lesson there are certain considerations that help you decide the best way to achieve the lesson objective. These considerations will largely determine the degree of student learning which will result from your instruction. Some of them will be your decisions, some will be determined by official directive. In making your estimate you will consider these major factors.

    b. <u>Subject Text</u>. This provides you with the essential information to be taught. Plan preparation, rehearsal, and presentation and consider the relationship of this lesson with other lessons in the same POI. This directive serves as the foundation for all lesson planning.

    c. <u>Program of Instruction (POI)</u>. This discussion of lesson planning mentioned the document which is the source directive for all instruction - the program of instruction. The program of instruction establishes the subject and overall scope of the lesson. It should be the first reference.

    d. <u>Students</u>. Another consideration is the class to be taught. Novice shooters will not receive the same level of instruction as that given to advanced students. The methods of presentation, the speed with which material is presented, and the complexity or depth of subject coverage will vary considerably. Consider what this class has received prior to your instruction. You do this so that you can properly relate your instruction to previous instruction and avoid needless repetition. You can secure much of this information from studying the program of instruction.

    e. <u>Time Available to Prepare and Present</u>. How much time will the instructor have to prepare his period of instruction? This time will be divided into intermediate target dates. This will allow time for last minute revisions of training aids, handout materials, and final rehearsals to polish up rough spots. Allocate time for your detailed coordination, and for preparing the completed vault file. Another important consideration is the number of hours allocated to present the instruction. This will influence the amount of information you can present, how deeply you will go into the subject and the degree of student proficiency you can reasonably expect. In a compressed marksmanship training schedule your task will be to insure student understanding within the time available.

    f. <u>Equipment, Facilities, and Personnel</u>. In planning the lesson you must consider the type of instructional area required and make a reconnaissance of this instructional area prior to completing your plans. By doing this you can insure that the instructional area will support the type of lesson and methods of instruction you plan to use. The availability of training aids and assistant instructors will also be important considerations.

2. LESSON OBJECTIVE.

    a. The lesson objective is a brief statement of the student learning to be achieved as a result of this period of instruction.

    b. <u>Necessary Elements of the Lesson Objective</u>. To be a workable guide for the instructor, the lesson objective must contain these elements:

        (1) The students who will receive the instruction; designate these students by identifying them by the specific course they are attending.

        (2) The degree or level of student proficiency expected as a result of this instruction.

(3) Subject. Here it is necessary to be specific. Pinpoint the portions of the subject you will teach.

(4) Learning to be Achieved. Specify the knowledge, appreciation, skills that the student should have learned and be able to perform as a result of this instruction. These learning outcomes are a further amplification or explanation of (2) above in concrete terms.

c. <u>Levels of Student Proficiency</u>. There are three levels of student proficiency. They are:

(1) <u>General Knowledge</u> - That level which makes the student aware of a subject and its general application to a major field. It provides the student with knowledge of the existence of certain fundamental facts and principles in a degree sufficient to enable him to recognize their implications and where to locate further information when the need arises.

(2) <u>Working Knowledge</u> - That level which makes the student sufficiently familiar with the primary purposes, major functions, or principles of employment of a subject to permit routine practical applications. It provides the student with sufficient knowledge and skill for him to apply without further training or experience. In addition, the student acquires limited proficiency to supervise or train others in that skill. It must be augmented by on-the-job training or further schooling for full qualification.

(3) <u>Qualified</u> - That level which provides the student with a comprehensive knowledge of the subject that permits skilled performance.

d. <u>Selecting a Desired Level of Student Proficiency</u>. You should consider these factors:

(1) The nature of the subject: Is it simple or complex? Is this the first formal instruction the student receives in this subject area or does he have previous instruction or on-the-job training? What is the training purpose as stated in the particular annex of the POI? Will this instruction be an orientation or are the students expected to become technically proficient?

(2) Time available to present the subject: How many hours have been allocated in the POI to present the subject?

(3) Student backgrounds: What is the educational background and experience of the students? The instruction must be so presented to enable the average student in the class to comprehend, not to bore the more experienced, and to satisfy the needs of the less experienced. (See Annex A for sample lesson objective.)

3. <u>STUDENT PERFORMANCE OBJECTIVES</u>.

a. <u>Explanation</u>. Once the lesson objective is formulated, you are ready to consider what must be taught in order to accomplish this objective. In other words, certain key facts or elements of knowledge must be understood. Military and civilian schools call them specific objectives, main thoughts, ideas or points, student learning outcomes, or primary areas to be taught or discussed. Once the student understands and can use these ideas the lesson objective has been accomplished. We shall call these elements of knowledge student performance objectives. Each must be a necessary and critical part of the overall lesson objective. The sum of the student performance objectives constitute the lesson objective.

b. <u>Definition</u>. A student ferformance objective is a statement in complete sentence form of a specific and significant principle, item of doctrine technique, skill, or element of knowledge that students must understand and be able to apply as a result of a period of instruction.

    c. <u>How Many Student Performance Objectives?</u> The instructor preparing a lesson is always concerned with how many student performance objectives there should be in any given lesson. The decision should never be based solely on number. Rather, select them by considering how many important parts or elements will be needed to accomplish the objective. There could be as many as ten sub-objectives in a two-hour problem or only one in an eight-hour problem. In addition to consideration of the elements of knowledge needed to teach the objective, there are other aspects of the problem to consider in selecting student objectives. Certainly the degree of learning dictated by the lesson objective will have a bearing on the number as well as the complexity of the performance objectives. Whether the problem is complete in itself or one of a series of integrated problems will be considered. Since student performance objectives are the heart of the lesson, give them much thought before selecting each one. Check carefully to insure that each is clear, concise statement of fact. Of course, the proof of each performance objective will be developed in the actual presentation of your supporting material.

    d. <u>How to State Performance Objectives</u>. State them clearly and briefly. Make sure they are complete sentences. Much care and work will be necessary to make them simple and brief, yet meaningful. A performance objective doesn't have to carry the whole burden of significance, however. Sometimes the supporting material will provide the critical backup. Watch out that you do not elevate supporting material to the level of an objective. Frequently a new instructor prepares a lesson with eight or ten sub-objectives. After he studies it carefully, he discovers that several of these so-called points are really valuable supporting material for one overall teaching objective. By further thinking through his problem, he discovers that he really has only three or four main points. Remember, a student performance objective is a significant and critical element of knowledge - refined and distilled. It is the product of prolonged careful thinking and considered judgment. (See Annex A for sample student performance objectives.)

   4. <u>SUPPORTING MATERIAL</u>. After your performance objectives have been established and listed in the most meaningful sequence you must then decide how you can best achieve understanding of these points. You do this by the use of good supporting material. We might compare student performance objectives and supporting material to the presentation of a law case by a lawyer. The lawyer states in his introductory remarks certain elements or facts which he will prove as the trial develops. These elements or facts correspond to our sub-objectives. Throughout the presentation of his case he proves or creates an understanding of each fact previously presented. This would be supporting material. So you too, in deciding upon how to prove or teach your points, must consider what will be the most efficient manner of presenting your case. The common ways are: by explanation, demonstration, discussion, skits, use of historical examples, quotations from famous people, practice exercises, and the use of visual aids (see Figure 3-2). As a well prepared instructor, you have a number of varied illustrations, examples, and restatements in reserve. Which of these you will use in supporting each performance objective will depend upon the degree of understanding desired, the amount of time you have to create this understanding, and the complexity of the point you are supporting. Don't over-teach or over-explain. You should insure that you have enough supporting material to explain the specific objective completely and adequately. How much supporting material you use will be determined by how soon most of your class grasps the projected point. This you can evaluate by using checkup questions. When satisfied that the class understands the first objective, proceed to the next. If the students do not grasp the point, restate, and supply additional example or illustrations until learning occurs.

   B. <u>ORGANIZATION OF THE LESSON</u>. Up to this point in lesson planning, you have studied the problem directive, examined the program of instruction, determined the lesson objective, and carefully selected the student performance objectives and appropriate supporting material.

1. Determine how to organize this material to insure effective student learning. Check these points:

  a. Will the sub-objectives selected provide the student with the knowledge essential to accomplish the lesson objective?

  b. Can these points be further consolidated or clarified?

  c. Are the performance objectives arranged in a logical, easy-to-understand order that will enhance student learning? Are they arranged from the simple to the complex, known to unknown, easy to difficult?

  d. Is the supporting material sufficient to prove the point? Have I used various types of supporting material to maintain student interest?

  e. What method(s) will I use to present the student performance objectives? (See Chapter 2)

Figure 3-2. Ways to Support Student Performance Objectives.

2. LESSON OUTLINE.

    a. The lesson outline indicates what is taught, in what order or sequence it is taught, and what teaching methods are used. The outline insures that teaching is properly planned and not haphazard or impromptu. Proper organization helps to ensure effective presentation.

    b. You prepare a full sentence outline for each problem you present. This lesson outline will follow the prescribed format as illustrated in the Sample Lesson Outline (See Annex A) and will be filed in the Lesson Outline Section of the Vault File for that problem.

    c. You do not teach directly from the lesson outline. After completing the outline yoy may use it for the first or second rehearsal. Then you prepare a set of teaching notes to keep on the lectern during the presentation. These notes are prepared in the way most convient for your use. They may contain phrases, abbreviations, color coded key ideas, or whatever symbols you wish to assist you during your presentation.

    d. This outline should be complete enough so that another instructor could prepare his own teaching notes or teach the subject if the emergency arose which precluded his doing the necessary original research prior to presenting.

    e. The heart of the lesson outline consists of the introduction, the body, and the conclusion. In preparing the lesson outline you will normally prepare the body first. This is the principal part of the lesson because your first decision was what must be taught to accomplish the lesson objective. When you have decided this, you are ready to prepare a proper introduction. As a final step you write the conclusion.

3. THE BODY OF THE LESSON OUTLINE.

    a. The body is that portion of the lesson outline where you develop carefully and completely the material to be taught. The body contains the student performance objectives together with the supporting material, lead-off questions with anticipated student responses, transitional sentences or paragraphs between major divisions, notes indicating where and when training aids are used, and planned sub-summaries.

    b. The first step in preparing the body is to place the sub-objectives and the material supporting each point in a logical order. Frequently the sequence of presentation is obvious; however, it may be necessary sometimes to experiment with the sequence before you actually decide the order in which you will present the various points. The two principal orders of presentation are the inductive or deductive order. In the inductive order you start with the individual specific examples or cases and build up to a conclusion or principle based on these examples. In the deductive order you state the basic principle or rule and proceed to prove or illustrate it by a series of examples. Remember this: if the order of presentation is not too clear to you it will completely confuse your students!

4. THE LESSON INTRODUCTION.

    a. The lesson introduction consists of four mandatory elements: gain attention step, lesson tie-in, motivation, and scope. Two other elements may be included when they are appropriate; student application of this instruction and the methods of instruction which will be used.

    b. <u>Gain Attention Step.</u>

(1) Gaining attention may suggest to some persons the use of a gimmick, a trick, or the use of a startling device. If so, the real need for and purpose of an attention-gaining step is misunderstood. Before you can expect to have students learning, or even to stimulate student desire for learning, you must have the attention of the student. You do this by gaining attention - focusing of student attention upon the subject to be learned, and not distracting students by a novel gimmick.

(2) Some ways to gain attention are to tell a story (humorous or historical), ask a rhetorical question, present a skit or conduct a demonstration, or merely walk to the center of the platform, pause until things are quiet, and commence your instruction. If the latter is a method of gaining attention, then we might ask ourselves, "Why is there a need for the other methods mentioned?"

(3) Sometimes you will be faced with the problem of presenting introductory periods or perhaps you are scheduled for the first period after a holiday or the first period in the afternoon of a very hot day. These instances are not extreme. You must be sure that the attention of the student is on you and on the subject which you are to present and not on some personal problem or activity. The method which you choose to gain attention will be selected only after a careful consideration of the class, the subject and the time allowed.

c. <u>Lesson Tie-In</u>. In the lesson tie-in you show the relationship of the problem you are presenting to other instructional problems which the students have received or will receive. It is easier for students to understand material if they can see the relationship of the present lesson to the overall block of instruction which they will receive. The fact that a problem is an introductory problem does not dismiss the need for a lesson tie-in. An accepted procedure for an introductory problem is to show the students the relationship of problems which they will receive in this instructional block to the accomplishment of the overall course objectives.

d. <u>Motivation</u>.

(1) Another mandatory element in the introduction is student motivation. This all-important step cannot be slighted. In stimulating your students, it is not necessary to "fire" them up as some people might think, but it is absolutely necessary to insure that the students know why they are receiving this instruction. They must appreciate the importance of the material being presented. You will never be assigned to teach a problem that is not important; unimportant problems are not presented. The importance of this subject then, must be communicated to the class.

(2) One of the ways in which you can arouse motivation is to relate this lesson to the skill which the student will eventually perform. In doing this, however, you must be careful not to cite absurd situations.

(3) Another method of motivating is to appeal to the pride of the class or make reference to the competitive spirit that this class should possess. You can also arouse motivation by relating this instruction to goals which appeal to the personal and professional ambitions which virtually all possess.

(4) A type of fear motivation (which may not stimulate long-term retention) is the statement that the students will be held responsible for this material on an examination.

e. <u>Scope</u>.

(1) A mandatory element of the introduction is the statement of the scope. Here you tell the students specifically what you and they are going to accomplish during this period.

You inform them of the learning which they should acquire during the period. This statement of the scope will come from the lesson objective which you selected when you initiated your lesson planning, but it will be more specific and developed in more detail than was that objective.

(2) After you have stated and discussed the scope, the student should have no questions as to what will be accomplished during the ensuing period. This discussion of the lesson objective(s) is a very important part of the general student orientation for the period and must not be hurried. Here you explain how the objective will be accomplished by giving the students an overview of what points will be covered in this lesson. The students can adjust mentally to how the lesson will be developed and thus be able to follow more easily.

(3) Some educators call this "the whole-part-whole method." In the introduction the student sees the big picture, then you develop the parts which make up that big picture in the body. Once again you wrap up these parts with another (but now a fuller) understanding of the "whole" as the lesson conclusion.

f. Sequence of Presentation.

(1) Any or all of the elements in the introduction may be combined. You are not restricted to presenting these parts in the order in which we have discussed them. Of course, you must gain attention before you present the other elements. Some instructors prefer to develop the motivation step immediately after gaining attention. They argue that it is good to motivate while interest is high. Present them as you will all of your instruction--in the most logical sequence, the way that seems best to you.

(2) Whichever sequence you employ, you must insure that before you go into the body of your presentation, the students understand what instruction is to be presented, how it ties into their instructional program, and why they personally need to know this. These are the three mandatory parts of the orientation.

g. Optional Portion of the Introduction. An additional element of the introduction which should be used when appropriate is explanation of methods of instruction. Orient students on the methods of instruction only when multiple methods will be used either in this problem or in the block of instruction. For example, if you plan to use a training conference during a portion of the problem and then use a demonstration during another portion and finally to require the students to perform, this should be announced to the students. This helps to get your students mentally set for what will follow. They will listen more carefully if they realize that they will be performing later in the problem.

5. PREPARING THE CONCLUSION.

a. The final part of the instructional problem is the lesson conclusion. This part of the period is unfortunately the part which is most frequently slighted. It is slighted because beginning instructors do not understand the need for an effective conclusion. They do not understand the real purpose of the conclusion. The conclusion, when properly presented, will point out the importance of the material which the students have learned and re-emphasize the main points. It will leave the students with the big picture of how these points all fit together to fulfill the lesson objective and what main ideas they should take away with them.

b. There are four parts to an effective conclusion: Retain student attention, summarize, point out application, and make strong closing statement. You do not need to present them in the order in which we will discuss them; you may combine and present them in the most logical or meaningful sequence.

c. <u>Retain Student Attention</u>. When moving into your conclusion, be sure that you have the full attention of the students. This is the first step of the conclusion. A common place for interest to sag is near the end of the period. Students know when a period or problem is about to be concluded and unless you insure that you have their attention, the effectiveness of your conclusion may be lost. One of the ways to maintain attention is by the effective use of a transition which will be discussed in some detail in Chapter 6. Other ways of maintaining or raising student attention have been discussed in the paragraph on the introduction.

d. <u>Summarize</u>. When you are sure that you have retained or stimulated the attention of the class, you then re-emphasize or summarize the main points of your lesson. Your statements or restatement of performance objectives, together with carefully selected key supporting material, constitute the lesson summary. Point out each main idea in the summary. Remarks such as: "in conclusion," "during this period we have covered," "let me see what we have discussed," "what have we learned today," are statements which through excessive use generally cause a loss of student interest. In approaching your summary, treat it as the third essential part of a good lesson rather than as a windup or just a way of dismissing your class. A good summary should challenge student imagination to use the main ideas presented in the problem; it is the climax of the learning situation. Effective learning must change student behavior.

e. <u>Point Out Application.</u> In this portion of the conclusion show students when, where, and how they will use the materials learned. If appropriate, state the immediate student application of these principles during their course and also the future application. The application step in the conclusion closely parallels the motivation step in the lesson introduction. Then you mentioned <u>why</u> it was important to learn this material; now, in the conclusion you show <u>how</u> students will apply what they have learned.

f. <u>Strong Closing Statement</u>. Finally, prepare a strong closing statement. Carefully plan this statement to stress again the importance of the material which the students have learned. Rehearse this closing remark until you can deliver it with maximum effectiveness. Remember the final impression you make upon your students is as critical as the initial impression. It is your final chance to drive home the importance of this learning. Make the most of it. Many good instructors make the final statement "the action step," what students should do about the principles or techniques they have just learned. Some conclude the lesson with an impressive demonstration. Here is a golden opportunity to impress the student with the importance of what you have taught him during the period. While you have his interest, drive home your message!

6. <u>COMMON ERRORS IN LESSON OUTLINING.</u>

a. New instructors inexperienced in lesson outlining should be aware of the following common errors. Use this as a checklist to review and revise your draft outline.

b. General:

(1) Instructors fail to use the approved lesson outline format properly.

(2) They omit the necessary indentations.

(3) They fail to insert the necessary note on when the handout is issued or when a specific training aid is used. Sometimes they insert the note but do it incorrectly.

(4) They fail to write the planned questions at the appropriate places in the outline or fail to write out the expected student responses as supporting material.

c. __Lesson Objective:__

(1) Frequently, the lesson objective is vague or too broad. The "what to teach" is so big or general that it cannot be taught in the time allotted. Actually the instructor plans to teach only a portion of the objective in that period. He states his objective as a "working knowledge of the pistol" when he means "a working knowledge __of the cycle of operation of the pistol cal . 45 (service).__

(2) Instructors frequently omit one of the three necessary elements of the objective.

(3) Frequently, the expected level of proficiency is too high for the training period. For example, it may be "to qualify" and the instructor allotted only a fifty-minute period to this qualification training.

d. __Lesson Introduction.__

(1) The introduction may be too vague, too brief.

(2) Frequently the __attention gaining__ portion is not written out. Sometimes there is only a brief reference. Remember, the outline should be complete enough so that another instructor could intelligently use it if such an emergency arose.

(3) Sometimes motivation is incomplete or not slanted specifically to the needs of this class.

(4) The __scope__ may not be specific enough to orient students so they can get mentally set for the __instruction__ to come. New instructors frequently think that the scope is simply a statement of the student performance objectives.

(5) The __transition into the body__ is usually too abrupt or sketchy. This transition is very important since it should smoothly set the stage for absorption of the first student performance objective.

e. __Student Performance Objectives.__

(1) Frequently the performance objectives selected are not the major items needed to accomplish the planned lesson objective.

(2) They are incompletely stated.

(3) They are vaguely stated.

(4) They are not in complete sentence form - merely phrases.

(5) They do not pertain to the objective as it is stated.

(6) They are not repeated in the lesson body at the appropriate place as the outline.

(7) They are actually supporting material for another objective and should be so subordinated.

f. __Supporting Material.__

(1) It may be too brief to substantiate the sub-objective or insure student understanding of the point advanced.

(2) It may be incomplete with key ideas assumed or inferred. Another instructor would not understand the development. Remember, you should have much more supporting material in the outline than you use during the presentation. These are the examples in reserve in case this class does not grasp the point as presented initially.

(3) The supporing material may be unrelated to the objective to be learned. Perhaps the relationship was not shown by a good transition.

g. Subsummaries.

(1) Frequently, several extensive teaching points are developed without any planned subsummaries.

(2) These subsummaries are not adequately written out.

h. Conclusion.

(1) The instructor sometimes neglects to plan for a good attention-retaining step. He doesn't use the same degree of imagaination as he did to gain attention initially. He fails to spend as much thought and effort in planning this step.

(2) The application step may not be specifically slanted to this class. Sometimes the instructor simply says, "Remember this and apply it when you get on the range."

(3) The summary may be too general. The instructor fails to reteach briefly the individual performance objectives and key supporting material. It may be too brief because he fails to allow sufficient time. He may overemphasize the later points and fail to mention the early ones.

(4) The closing statement may be good but the instructor overelaborates on it, explaining it in more detail and thus weakening its effects. To be good, a closing statement should be short. Brevity is the soul of effectiveness.

7. INSTRUCTOR'S TEACHING NOTES.

a. You never teach directly from a lesson outline. In the first place it is too detailed to permit you to orient yourself rapidly if you lose your train of thought. Secondly, as a written document it contains more formal diction that you would use in speaking. The essence of good teaching is to speak in a way natural to you. Therefore, you use the lesson outline as a source for preparing a personal series of teaching notes. These will be on the lectern for ready reference.

b. Most instructors find that good teaching notes are an efficient aid to presenting instruction. Simply knowing that they are available builds confidence in the beginning instructor. Do not use them unless you have to, for you may develop the bad habit of using them as a crutch and never rehearsing your presentation sufficiently. These notes should be organized in such a way that you can easily present your instruction without awkward stops to relocate yourself in the lesson. Make them large enough to be read at a glance.

c. Other instructors use key words or phrases as notes to cue them into the main portions of their lesson. Place the notes in a spot where they are readily visible for your use, but do not distract students. Effective instructors color code the significant parts of their notes to indicate student objectives, training aids, handouts, or other presentation cues.

8. PREPARE INSTRUCTIONAL AIDS AND HANDOUTS. After the lesson has been planned you must decide the number of assistant instructors needed and then orient and rehearse them. If you are going to need advance sheets and/or new training aids, prepare for these materials early. You accomplish all this after your lesson has been planned. It is wise to prepare a checklist for your problem when you begin your lesson planning and as you accomplish each step in the overall planning of your problem, check off this item as completed.

9. REHEARSING.

a. Once you have planned and written out your lesson outline and established all of the problem requirements, you are ready to rehearse. The type of problem is a prime consideration in establishing your rehearsal schedule. It is possible to rehearse too much, just as it is possible to rehearse too little. The number of rehearsals which you conduct before presenting a problem will depend upon how well you have researched your problem and how familiar you are with the technique which you will employ.

b. As a minimum, three major rehearsals are required for each lesson. You should have a learning rehearsal which you conduct by yourself and for yourself. You should have a "murder" rehearsal during which all of your training aids, handouts, and assistant instructors are used. The murder rehearsal should be attended by your supervisor and selected team members. As a result of this rehearsal, you will receive comments and criticisms from experienced personnel and revise your problem accordingly. Finally, you should have a full dress rehearsal using every device, method, and technique just as you will use them during the actual conduct of the class. You may have to conduct more than three rehearsals but the decision depends on you. You alone know when you are ready to go. Knowing your material thoroughly will reduce apprehension and stage fright on the day of actual presentation. Rehearsals are the only way to check your knowledge of the subject and build up and retain self-confidence.

ANNEX TO CHAPTER III, "LESSON PLANNING".
(Sample of Lesson Plan for Principal Instructors)

UNITED STATES ARMY MARKSMANSHIP TRAINING UNIT
Office of the Director of the Small Arms Firing School
Fort Benning, Georgia

EFFECTIVE SPEAKING (I-4)   Apr 65

LESSON OUTLINE

I. LESSON OBJECTIVE: To enable Advanced Pistol Marksmanship Instructor Training Course students to explain how to use the voice and body effectively to communicate and reinforce meaning.

II. STUDENT PERFORMANCE OBJECTIVES: As a result of this instruction the student must be able to:

    A. DEFINE and EXPLAIN speech as "the communication of thought and emotion by means of the voice and bodily action."

    B. EXPLAIN how the physical organs of the body produce speech and make hearing possible.

    C. EXPLAIN how proper use of the voice makes communication possible and helps to reinforce meaning and understanding.

    D. ILLUSTRATE how meaningful gestures, platform movement, and facial expression supplement the voice and add effectiveness to communication.

III. ADVANCE ASSIGNMENT: The Advanced Pistol Marksmanship Instructor's Manual, Vol II, Chapter 4, "Effective Speaking".

IV. INTRODUCTION: 3 Min.

    A. <u>Gain Attention:</u>

NOTE: PLAY TAPE EXAMPLES (EFFECTIVE AND INEFFECTIVE VOICES).

        Obviously each of us speaks with varying degrees of ability, depending upon the practice and training he has had.

    B. <u>Orient Students:</u>

        1. <u>Lesson Tie-In:</u> This hour of instruction is the climax of a two hour block of instruction on speech. During the first hour you considered the instructor alone. This hour we are going to discuss the actual process of making yourself understood through the use of effective speech.

        2. <u>Motivation:</u> In spite of the wonderful training aids we have today -- Overhead projectors, film projectors, tape recorders, black lights, magnetic boards, and models -- it is still the human voice and the process of speech that ties these things together and gives them meaning and purpose. No one is ever completely effective in communicating his thoughts to his listeners. We must, therefore, concentrate on using our voices as effectively as possible in the process of speech if we are going to insure maximum learning.

2. <u>Scope</u>: You learn to speak effectively by knowing:

    a. How to define speech - What it is and what it means.

    b. The physical process of speech and hearing.

    c. Factors that influence the voice and techniques of improving their use.

    d. How to reinforce the voice by proper use of gestures, movement and facial expressions.

<u>NOTE</u>: ASK IF THERE ARE ANY QUESTIONS ON THE ADVANCE ASSIGNMENT AND ANSWER THEM.

V. BODY:

    A. <u>First Student Performance Objective</u>: Students must be able to define and explain speech as "the communication of thought and emotion by means of the voice and bodily action."

        <u>QUESTION</u>: What is your definition of speech?

<u>NOTE</u>: SHOW SLIDE #1 (DEFINITION OF SPEECH).

        1. Speech is the communication of thought and emotion by means of the voice and bodily action.

        2. The entire being, to include the mind, voice, arms, hands, legs, and face, is needed if we are to speak effectively.

        3. Our definition of speech includes the word "communication". This is, in fact, the sole purpose of speech. Communication means mind-to-mind contact between the instructor and the student that allows information to pass from one to the other.

<u>TRANSITION</u>: Now that we understand the meaning of the word "speech", let's see just how it happens. Where did it originate? How does the body produce speech?

    B. <u>Second Student Performance Objective</u>: Students must be able to explain how the physical organs of the body produce speech and make hearing possible.   5 Min.

<u>NOTE</u>: SHOW SLIDE #2 (LUNGS, DIAPHRAGM, ETC.).

        1. Sound is produced when the chest muscles and diaphragm force air from the lungs up the windpipe and through the voice box. The vocal cords produce sound when they are vibrated by the air passing over them.

        2. This sound is molded and formed into words by the action of the jaw, tongue, teeth, and lips. These are called the articulators.

<u>TRANSITION</u>: That, briefly, is how our voice is produced. This voice produces sounds which are formed into words by the articulators. The combination of these words form ideas and thoughts over direction of the mind. This is how we produce a speech -- so far involving only the speaker. But true speech is part of communication, and communication requires receiving -- or hearing.

NOTE: SHOW SLIDE #3 ("EAR AND BRAIN").

       3. The listener's brain translates sound waves made by the voice into thoughts.

          a. Sound waves strike the eardrum and cause it to vibrate.

          b. This vibration is transmitted to the inner ear and creates nerve impulses which are then transmitted to the brain. Here they are translated into ideas and thoughts on the basis of previous knowledge and experience.

          c. The inner ear analyzes sound in three ways:

              (1) By energy: This is simply speaking loudly enough to be heard so that maximum learning takes place.

              (2) By rhythm: For our purposes, this involves using the proper rate of speech and making correct use of pauses, both of which will be discussed later in the period.

              (3) By pitch: The more varied the pitch of your speech, the more interesting your speech becomes. Thus, understanding occurs more readily.

       4. The ear is constantly receiving composite sounds.

          a. Even while sitting in a classroom listening to instruction, a student hears other sounds. He is conscious of some, oblivious to others. (Examples: passing vehicles or airplanes, voices outside, humming of the ventilation system, or high heels clicking on the sidewalk or in the hall.)

          b. As instructors, you must recognize composite sounds as a potentially serious distraction and develop your ability to use your voice to overcome these sounds and to hold the student's attention.

       QUESTION: How can you improve the effectiveness of your voice?

TRANSITION: By recognizing the factors which affect speech, learning how to take maximum advantage of these factors and by constant practice. The first factor is rate.

    C. <u>Third Student Performance Objective</u>: Students must be able to explain how the proper use of the voice makes communication possible and helps to reinforce meaning and understanding.   20 Min.

       QUESTION: What is the proper rate of speech in words per minute?

       1. The average rate of speech is about 125 words a minute.

       2. Rate varies among individuals and is greatly dependent upon the geographical region you come from.

       3. There is only one rule and that is a general one. Speak fast enough to be interesting, but slowly enough to be understood.

       4. A proper or interesting rate of speech:

          a. Insures understanding of your words, thus aiding the student in understanding your thoughts and ideas.

      b. Adds interest, primarily through a change of pace.

      c. Helps to place emphasis on important points - slow up for emphasis.

      d. Adds meaning by helping to create "word pictures."

NOTE: PLAY GODFREY TAPE OF FDR FUNERAL.

    QUESTION: How can you improve your rate of speech?

   5. Develop the habit of listening to yourself and constantly practice the following:

      a. Say what you mean! Don't let your speech get ahead of your thoughts.

      b. Say your words distinctly so that each one may be understood.

      c. Gear your rate to fit the mood and complexity of your ideas. Light unimportant thoughts are best expressed at a comparatively fast rate; serious ones more slowly. Complicated explanations require a fairly slow, methodical rate.

      d. Emphasize or reinforce meaning by slowing or speeding your rate.

TRANSITION: We have determined that rate should be governed, to a degree, by the meaning of the thoughts or ideas. Those words which stand for a single thought or idea are grouped together into phrases. In writing, this grouping is accomplished by punctuation marks; in speaking, it is done by the effective use of pauses.

   6. Pauses are the principal "punctuation marks" of speech. They separate groups of words into thoughts or ideas.

   7. Pauses accomplish four things:

      a. The listener gets a chance to mentally absorb what has been said.

      b. The speaker gets a chance to think of his next point.

      c. The speaker gets a chance to breathe.

      d. Emphasis and meaning are given to thoughts and ideas.

NOTE: PLAY ROOSEVELT TAPE.

   8. Read aloud and use pauses to help give the words their intended meaning.

   9. Do not be afraid of pauses. People who overuse "uh" and other meaningless sounds believe that they must be making sound even when they are not expressing ideas. Others are afraid of silence, as a child fears the dark, and hurry on to fill it with sound.

TRANSITION: Pauses are not the only punctuation marks of speech. There is also inflection which is the change in pitch of the voice as one speaks, the movement of the voice up and down the musical scale, or vocal variety.

   10. It is one of the most effective devices an instructor can use to sway the audience a monotonous voice puts them to sleep.

11. Inflection makes speech more interesting but also has a direct influence on the meaning.

NOTE: PLAY COMMENTATOR'S TAPE.

QUESTION: How can you improve your use of inflection?

12. Inflection is not mechanical. It cannot be added to one's speech just for the sake of adding it.

13. Inflection results directly from the speaker's feeling for his subject. It is, therefore, a product of sincerity and enthusiasm.

14. If you lack knowledge of your subject, you will lack confidence. This will cause you to worry about your known weaknesses, speak faster than normal, or you will stammer, and speak in a monotone.

15. Even with full knowledge and confidence, some speakers are ashamed to show their true feelings about a subject. They speak in a dignified manner but their voice lacks life, warmth, and conviction. This lack is quickly transmitted to the audience.

TRANSITION: Rate, pauses, and inflection make speech more interesting, and they add emphasis and meaning to your words. However, complete understanding will not be achieved unless your words are spoken clearly and correctly.

QUESTION: What is meant by pronunciation?

16. Pronunciation is the sounding of a word with the accent on the proper syllable or syllables, in accordance with a given standard (usually a dictionary).

17. The principal reason for poor, or incorrect pronunciation is carelessness -- not taking the time to look up the correct pronunciation of a word when you are in doubt.

18. Often, however, the pronunciation of certain words varies according to the section of the country in which one has spent a good deal of time.

19. The most important thing, of course, is to be understood. But many people are satisfied just to be understood, even though their pronunciation of many words is incorrect.

    a. We are not speaking now of five-syllable words that are seldom heard or used.

    b. Most errors in pronunciation are heard in little everyday words. For example: "winder" for window; "hep" for help; "jist" for just; "git" for get.

    c. If you are going to use big words, be sure you know the meaning and correct pronunciation before you do.

20. Correct pronunciation is not "sissified" nor is it the mark of a "stuffed shirt." It is an absolute necessity if you are going to be understood. Listen closely to people whose speech is cultivated, and learn from them how to get into the habit of pronouncing words correctly.

21. Finally, if you have an accent common to a particular part of the country, don't be ashamed of it or try to hide it. Use it! It's a mark of individuality and helps to make you more interesting to listen to -- but do attempt to improve it.

<u>TRANSITION</u>: Closely related to pronunciation, and actually a part of it, is the correct forming or shaping of words so that they are heard distinctly.

22. Articulation is the formation or joining of sounds to form words. Good articulation results in clear, distinct speech.

23. Poor articulation results in slurring, muffling, or mumbling of words. It is caused by failing to use the articulators to form parts of words -- in other words, just plain laziness.

24. Most people can correct sloppy articulation by applying themselves to the job of speaking clearly for the benefit of their listeners: listen to words, say each one as a conscious effort.

25. Be conscious of how you say words. Clear, distinct speech is something which marks you as a person who cares enough about others to make himself understood.

<u>TRANSITION</u>: Each of the factors we have discussed is a necessary part of clear, interesting speech. None, however, is more important than just being heard.

26. Volume in speech is technically defined as vocal energy. We know that all of you can be heard in a classroom or in front of a set of bleachers, so we all emphasize two other factors which depend on volume: force, and quality.

    a. First, when working with a microphone, you must acquire the ability to be completely oblivious of the mike and the sound system. They exist only to reproduce your voice and insure that it reaches your entire audience.

    b. Force is a part of enthusiasm. It is another way of giving emphasis to your ideas. It is simply "hitting" certain words harder -- saying them louder in order to stress their importance.

    c. A sound system will help you be heard, but it will not improve the quality or tone of your voice.

        (1) Some voices are pitched high, some medium, some low. But your natural pitch is the one you should use. Don't be phony or use a tone different from that which is natural for you.

        (2) Many persons are self-conscious about what they think is a high-pitched voice. There is no need for this. The important thing to remember about quality is:

            (a) Breathe deeply between ideas. Use your lungs to back up your words with plenty of breath pressure.

            (b) Open your mouth and let the words out--don't mumble them.

            (c) Avoid talking through your nose, thereby having a nasal quality or twang to your voice. Also, don't use a throaty or oral delivery. Reach a happy medium between the two.

TRANSITION: The factors of speech, then, are rate, pauses, inflection, pronunciation, articulation, and volume, all of which can be improved only by having the desire to improve. In simpler words, practice, but the voice isn't all. It isn't enough merely to be able to talk well.

    D. <u>Fourth Student Performance Objective</u>: Students must be able to illustrate how gestures, platform movements, and facial expressions supplement the voice and add to the effectiveness of communication.   15 Min.

        1. Gestures are movements, usually of the arms and hands, which serve to clarify and emphasize ideas.

        2. It has accurately been said of some persons that if their hands were tied, they couldn't say a word. But for some reason, most of us are reluctant to use gestures when speaking in front of a large group.

            a. Gestures must be natural. Therefore, they should not be practiced and used the same way every time one speaks.

            b. A thorough knowledge of your subject and a sincere desire to sell it to the students are the secrets to the use of effective gestures.

        3. Begin your speech with your hands and arms in any position you feel appropriate to what you are saying. Then let nature and your own enthusiasm take their course. One word of caution: Your gestures must be spontaneous.

        4. We will not go into gestures any further in this class. The group leaders will cover gestures very thoroughly in group work.

TRANSITION: Standing in one place during all or most of a presentation, even though you speak well and gesture effectively, would become tiring to your audience and to yourself.

        5. Movement on the platform must be natural and completely spontaneous. The moment you begin to move just to be moving, you become obviously stiff and unnatural. However, in rehearsals, it is wise to consciously try different types of movement just to get the feel of the platform.

            a. The principal advantage of movement is that it helps to hold the attention of the audience. The eye naturally follows a moving object and focuses on it.

            b. Transitions between points can be indicated and made more emphatic merely by shifting the weight from one foot to the other or by lateral movement of a step or two.

            c. Lateral movements should be started with the foot on the side toward which you are going. This avoids awkward crossing of the feet.

            d. A step or two forward might be used to indicate the importance of a point.

            e. Backward movement suggests that you are willing for your audience to relax a bit to let the last idea take root before you present another one.

            f. The basic rule to remember is moderation and naturalness. Don't stay glued to one spot or one small area, and don't be on a racetrack all of the time.

TRANSITION: We have discussed just about every moving part of the body but one has been left out.

  6. Facial expression is invaluable in reinforcing meaning and maintaining the proper relationship with your audience.

    a. There is no rule to follow. You must realize that a "poker face" does not belong on the platform. To avoid it, know your subject, organize it thoroughly, and devote yourself to convincing and teaching your students -- your face will take care of itself.

    b. The most common expression is one of serious concentration, resulting primarily from fear that part of the lesson will be forgotten and left. Know your material! Then you can concentrate on your students as people, and your face will reflect your interest in them.

VI. CONCLUSION: 5 Min.

  A. <u>Retain Attention</u>: Now you should be asking yourself one question: Can I teach? Your answers will probably vary, but also important is the answer to this question: Do you want to teach?

  B. <u>Summary</u>:

  1. Use your voice and body to communicate thought and emotion. To speak effectively, you must make use of your mind, voice, arms, hands, legs, and face; in short, you must bring your entire being into action.

  2. Sound is produced when the chest muscles and diaphragm force air from the lungs up the windpipe and through the voice box. This sound is formed into words by the articulators to include the jaw, tongue, teeth, and lips. The sound waves of the voice are translated into thoughts by the listener's brain. The inner ear analyzes sound in three ways: by energy, by rhythm, and by pitch.

  3. To insure the effectiveness of your voice, you must employ proper rate, pauses inflection, pronunciation, articulation, and volume.

  C. <u>Application</u>: You will have the opportunity to apply all you have learned in every period of instruction that you will present. Success in any field is more dependent on one's ability to express himself than any other single factor.

  D. <u>Closing Statement</u>: Everything we have discussed is common sense and certainly not new. We have given you no new rules nor have we tried to establish any kind of a false standard by which to measure your performance during the course. The key to successful, effective speaking is to know what is right, then do it -- each in your own way.

CHAPTER IV

EFFECTIVE SPEAKING

A. <u>GENERAL</u>. CLEAR COMMUNICATION IS ESSENTIAL. Why should you be concerned with your ability to speak? You've been speaking for years and have had no difficulty in making yourself heard. But being heard is not enough. No matter what your purpose in speaking, you, as a contributing member of society, want to make certain your thoughts and ideas are understood, not just heard. You want people to understand your words as the first step toward a richer goal: that your thoughts and ideas be believed, felt, learned, and remembered. All this is true of the person who realizes his social responsibility. But as a marksmanship instructor, it is not only desirable that you make people believe, feel, learn, remember and act, it is essential, it is your duty! You must make ideas vivid in your student's mind---so vivid that a chain reaction of new ideas is jarred loose, and the student takes your lead and thinks for himself. You must make him feel the power that new knowledge gives him, the power of competence and assured success. You must teach him what you know! And how will you do this? There is only one answer -- <u>by learning to communicate effectively</u>!

B. <u>YOUR VOICE</u>. VOICE - THE MASTER CONTROL OF SPEECH. Teachers of speech have compared the voice to an exceptional type of musical instrument. Like other instruments, it has pitch, range and volume. Like other instruments, it can communicate feeling as well as sound. With your voice you can convey emotions such as rage, reverence, amusement, love and seriousness. When voice techniques are appropriate for the words to be spoken, the combination produces a potent communications tool.

1. <u>ARTICULATION</u>.

    a. There are two basic types of speech sound: vowels (a, e, i, o, u, and sometimes, y) and consonants. <u>Articulation</u>, for our purposes, is defined as the production and combination of separate sounds to produce intelligible speech. It involves the action of the tongue, teeth, lower jaw, and lips on the breath stream to form specific sounds. You achieve loudness of speech through the vowel sounds; you achieve intelligibility mainly through the consonant sounds. Hence, the maxim: "The vowels give beauty, the consonants give clarity." Say the vowels to yourself and notice that the breath stream is <u>shaped</u> to obtain the difference in sound, not <u>interrupted</u>. Now say the consonants and notice that each one involves constriction of the breath stream sometimes to the point of interruption by the tongue, lips, or throat muscles.

    b. To be a good speaker you must have as your goal distinctness in articulation. You must avoid slurring, mumbling words. With the exception of mental speech difficulties such as cleft palate or stuttering, there is only one common cause of poor articulation: Laziness!

    c. Practice saying words as you know they should be said. "Explode" the <u>p</u> in pull, the <u>t</u> in talk, and carry this through to your everyday speech. Relax your throat, tongue, jaw and lips and use them to clip off crisp sounds and blend sounds into clean, bright, clear words. Don't be lip lazy!

2. <u>PRONUNCIATION</u>.

    a. Pronunciation is the sounding, or articulating, of a word with the accent on the correct syllable in accordance with good English usage. The principal difficulties associated with pronunciation arise from sheer ignorance and distinctly regional accents. Ignorance of correct pronunciation can be overcome by acquiring the habit of listening carefully to cultivated speakers and by using the dictionary when in doubt. If you say <u>jist</u>, <u>git</u>, <u>gunna</u>, <u>what</u>, <u>whatcha</u>, hafta, or any of the many other commonly heard "easy" ways of saying words, your speech is sloppy and you are faced with a cleanup job. Just pay attention to what you're saying, as you expect other people to do; remember, they won't if you don't

    b. If you have a regional accent, such as a Southern drawl or a New England twang, don't try to eliminate it - make the most of it! It's part of your personality. Just don't let it get out of hand to the extent that people from other parts of the country can't understand you. A small degree of accent is pleasant to the ear and adds interest and personality to your speech.

  3. GRAMMAR.

    a. Grammar is concerned with the correct usage of the spoken or written word. A sense of rightness or wrongness of grammar comes more from familiarity with good literature and association with well-spoken people rather than from memorization of academic rules and principles. Above all, you must _want_ to speak correctly. You must be mentally alert to what you are saying and make conscious and continual effort to maintain an acceptable level of grammar. Association with persons using poor grammar insidiously affects your speech unless you guard constantly against the tendency to slip into careless grammar.

    b. Glaring errors, such as "him and me is going", "I seen", "he give" and "it run should be attacked immediately and corrected. But good spoken grammar must be based primarily on a long-range program. You must acquire a continuous consciousness of right and wrong usage, not for a period of weeks or months, but for a lifetime. Read as much as you can of the world's good writing, listen closely to people whose speech is cultivated and make use of books on grammar and correct usage to answer any special questions which arise. Then practice good grammar. Your speech, and your personality, will show a definite change for the better.

  4. RATE. Rate of speech is the speaker's speed of delivery. Each person has his own "normal" rate on conversational speech. This is, to a large extent, a result of nationality and geographic background - the environment in which you grew up, the section of the country where you spent the most time. There is no standard, proper rate of speech. You can't afford to speak at a slow, plodding rate that puts your listeners to sleep; neither can you rattle off words so rapidly that they run together and cannot be understood. As a general rule, speak fast enough to be interesting, slow enough to be understood. Just as a good baseball pitcher keeps the batter on his toes with a slow ball or a fast ball, take advantage of a vocal "change of pace" to hold the interest of your audience. Let your rate of speech be governed by the complexity of the thought, idea, or emotion you are trying to communicate. You can give motion to word-pictures by a rate of speech appropriate to the picture described. Use a fast rate for joy, excitement, vigorous action. Depending on your personality and the idea you wish to give to your listeners, add emphasis by either slowing or speeding your rate.

  5. COMMON RATE DIFFICULTIES AND SUGGESTIONS FOR IMPROVEMENT.

    a. Slow, Ponderous Rate. Force yourself to think faster so that you may also force yourself to speak faster. Plan to express ideas rapidly or slowly, as suits the mood of the idea. Read aloud and interpret the meaning of the words by the rate at which you speak them. Use a recorder to listen to your rate and then do it over to cut down the total time.

    b. Fast, "Machinegun" Delivery. Curb your impatience to blurt out your ideas. Take time to be clear. Force yourself to slow down. Recognize the listener's need to absorb your ideas, and give him time to do so by saying words clearly and pausing longer between ideas to let the ideas "sink in". Read aloud and observe the marks of punctuation. Express the meaning of the words carefully at the rate which fits your interpretation. Taking care to enunciate more precisely will generally slow your rate.

c. Halting, "Choppy" Rate. Concentrate on speaking complete ideas or sentences. Take a deep breath before you begin a sentence and do your breathing between, not in the middle of, ideas or phrases. Sometimes a choppy rate is the result of tenseness or nervousness. Work off excessive energy through physical activity just prior to talking or by exchanging pleasantries or comments with persons around you before you begin your presentation.

6. PAUSES.

a. In writing, we use periods, commas, question marks, and other punctuation marks to separate thoughts and ideas and to give the desired meaning and emphasis to our words. In speaking, the same functions are accomplished, to a large degree, by pauses. Pauses are also used to gain certain effects: humorous, dramatic, thought-provoking. Remember that pauses are a means of punctuation for effect.

b. The proper use of pauses accomplishes four things: Listeners are able to absorb ideas more easily; you get a chance to concentrate on your next point; you give emphasis, meaning, and interpretation to your ideas; and you get a chance to breathe!

7. COMMON PAUSING DIFFICULTIES AND SUGGESTIONS FOR IMPROVEMENT.

a. Not Enough Pauses. Begin by reading aloud something you like and force yourself to pause between ideas, at periods, commas, and other punctuation marks. Try to adapt the attitude of the artist who makes a few brush strokes then steps back to evaluate the results.

b. Too Many Pauses. This difficulty is caused usually by a lack of knowledge of the subject, failure to organize material thoroughly, or inadequate rehearsals. The speaker may not be certain of the sequence of his ideas and so must take the time to organize them while on the platform. Study your material, organize it on paper, then rehearse until your thoughts and words flow smoothly. Thorough familiarity with the subject matter increases personal fluency.

c. Over-Use of "Uh". Don't be afraid to pause! Silence, properly placed in the flow of speech, is more effective than any number of words; it certainly is more effective than meaningless, gutteral sounds. Use the same techniques suggested for b (Too Many Pauses) and leave out the "uh". Snip off these "speech whiskers" to improve your effectiveness.

d. Pausing in the Middle of Ideas. Here, your problem is to coordinate two necessary functions: thinking and breathing. Think of your listener as you speak; concentrate on making each idea clear by pausing only between ideas. Then, coordinate your breathing with the phrasing of your words - short pauses between phrases, deep breaths or long pauses between ideas.

8. INFLECTION.

a. Just as musical notes become melody when they are arranged in different relative positions on the musical scale, your voice becomes more interesting and your words more meaningful when you make use of changes in pitch; this is inflection, or "vocal variety".

b. Here is an example to illustrate how inflection on different words change the meaning of a question. Say the question to yourself, raising your pitch on the underlined words, as indicated:

Figure 4-1. Inflection - Key to Expression.

c. Inflection is the master key to expression of all kinds - emotional, persuasive, convincing. With it, you can move an audience to tears or laughter, imprint your ideas indelibly on their minds; without it, you will put them to sleep.

d. Like the pause, inflection is a way of punctuating speech; it puts the question mark at the end of a question, makes a statement of fact more positive, and helps to put an exclamation mark at the end of a strong statement. Inflection is the principal difference between just saying words and speaking ideas with meaning.

9. <u>COMMON INFLECTION DIFFICULTIES AND SUGGESTIONS FOR IMPROVEMENT.</u>

    a. <u>Monotone (No Inflection)</u>. We have said that inflection conveys feeling and meaning. But this is a situation similar to the old problem of which came first, the chicken or the egg, because feeling produces inflection. You must be willing to show your feeling about what you say. This is what is meant by personalizing your speech. Read aloud to a listener and practice using inflection to show him the meaning of the words. First, analyze "What is the emotion or feeling of this selection", then, convince yourself of the necessity for you to communicate this emotion through your inflection. A tape recorder is an excellent device to improve your inflection because here you must communicate emotion entirely through the voice; there are no gestures or visible expressions to help.

    b. <u>Misplaced Inflections</u>. Generally speaking, use downward inflection at the end of sentences to express positiveness and conviction. Be careful of downward inflection within the sentence itself, for each downward inflection gives a sense of finality to the thought phrase, and creates a mental break in the listener's thoughts. Use slight upward inflection within the sentence to indicate the thought is not yet complete; this serves to bind ideas together and to give unity to the thought. Give an upward inflection at the end of sentences only when you are implying question or uncertainty.

10. <u>FORCE.</u>

    a. Forceful speech combines the volume of carrying power of the voice with the demonstrated vitality and strength of conviction of the speaker; it includes the proper placement of stress or emphasis on key words and phrases. Like rate, pauses, and inflection, force is a way of conveying feeling, giving meaning, or adding emphasis. Yet, unlike the three factors previously considered, it cannot be set apart distinctly. It involves rate, pauses, and inflection plus carrying power, fullness of tone (or body), and proper regulation of loudness.

      b. Listeners will not respond properly to constant banging of their ears by a speaker who shouts and is insensitive to their feelings. Neither will they be convinced by the cool, detached manner conveyed by a speaker who is consistently calm, quiet, conversational, or patronizing. In order to communicate, we must awaken reactions and feelings in our listeners.

      c. Knowledge of subject and a firm grasp of the sequence in which you plan to present ideas will enable you to project yourself into good mental contact with your audience, leading their thoughts now by calmness, then driving home a point with power, now letting pure silence underline the significance of your words.

11. <u>COMMON DIFFICULTIES WITH FORCE AND SUGGESTIONS FOR IMPROVEMENT</u>.

      a. <u>Lack of Volume</u>. Concentrate on making the person farthest away hear you. Practice in a classroom and convey your meaning to an imaginary someone seated in the last row of seats. Most people can obtain this carrying power by raising the pitch of their voice and adding more than their normal amount of nasal quality. Above all, do not shout! Do not force your voice. Keep your chest and throat relaxed because if your throat muscles tighten too much your pitch will rise too high. Take a deep breath and sound off. Volume comes from the diaphragm not the throat.

      b. <u>Dropping Volume at End of Words or Sentences</u>. This is usually the result of incorrectly associating a drop in volume with downward inflection. Develop the habit of paying attention to the sound of your own voice and you will be able to judge whether you are being heard. Practice lowering the pitch of your voice without dropping the volume. Record your voice so that you can sit back and analyze how you sound to others. Read aloud, and concentrate on projecting every word in a thought or idea to an imaginary listener seated in the rear of the room.

      c. <u>Failure to Give Emphasis to Main Points and/or Key Words</u>. Know your subject! Identify your main points; then practice putting them across with a spurt of energy in your voice. You can't expect your listeners to do the work of sorting out the main ideas from the subordinate ideas. You must interpret for them by stressing key words and phrases with volume, pitch, rate, and pauses.

12. <u>GENERAL RECOMMENDATION FOR SPEECH IMPROVEMENT</u>. Although various specific means of correcting common difficulties have been indicated in this section, there are four ways in which you can improve all aspects of your speaking voice:

      a. First, use a tape recorder periodically to hear how you sound to others. This will enable you to make on-the-spot corrections of errors which otherwise would have gone unnoticed.

      b. Secondly, read aloud to one or more persons. If you have children, or a patient wife, or both, they will make a good audience. Choose selections you particularly enjoy, interpret the meaning and communicate this meaning by the way you use your voice.

      c. Thirdly, listen closely to polished speakers on radio and television (Edward R. Murrow, Eric Sevareid, Douglas Edwards, Howard K. Smith, Walter Cronkite, Chet Huntley). Do not try to make your voice or manner of speaking exactly like that of one of these men, but notice how they use their voices to give meaning to their words and emphasis to their ideas.

d. Finally, listen to yourself! Acquire the habit of evaluating constantly how you use the speech factors listed in this section and you cannot help but improve your effectiveness.

C. <u>YOUR BODY</u>. THE BODY SPEAKS. With improved speech comes the realization that the voice alone is not enough. If voice communicated perfectly, television never would have become popular. Seeing a person as he speaks adds immeasurably to the listener's appreciation of and receptiveness to ideas, IF the speaker uses his body to reinforce, emphasize, and clarify his ideas as he speaks then. Skilled pantomime actors are able to communicate every human emotion through the clever use of body, gestures and facial expressions alone.

Figure 4-2. The Body Speaks.

1. <u>POSTURE</u>. Posture is the speaker's stance. Your position should be comfortable without being slouchy, erect without being stiff. Let the weight of your body rest on the balls of the feet rather than on the heels. Stand erect with the assurance of command! The way you stand is an outward manifestation of an inner attitude. It will definitely affect the way your students receive your instruction.

2. <u>MOVEMENT</u>.

a. Movement is the motion of the whole body as it travels about the presentation area. One effect of movement is to attract the attention of the listener; the eye instinctively follows moving objects and focuses on them. Movement can greatly assist in conveying the thought of the speaker. Transitions from one point in the speech to another often can be indicated and made emphatic merely by shifting the weight from one foot to the other or by lateral movement of a step or two. Such a movement is an informal signal that "I am finished with that point; now let's turn our attention to another". Always start lateral movements with the foot on the side toward which you are going; this avoids awkward crossing of the feet. Then walk a step or two, naturally, in that direction. Forward and backward movements imply the degree of importance of an idea. A step forward implies that you are coming to an important point which you don't want your audience to miss; this emphasizes the point. A backward step suggests that you are willing for your listeners to relax and let the last idea take root before you go on to the next one.

b. The basic rule is moderation: Don't remain glued to one spot, and don't keep on the move all the time. When you do move, move briskly and with purpose. As your skill and experience increase, you will find your movement becoming less obvious and more meaningful, and you will learn to modify the degree of movement to make it natural and meaningful.

Figure 4-3. Improper Platform Behavior.

3. <u>GESTURE</u>.

    a. Definition. A gesture is a natural movement of any part of the body to convey a thought or emotion, or to reinforce oral expression. Your arms, hands, and body, are your principal tools of gesture. Practice gestures as natural parts of your speaking manner, but never rehearse specific gestures to use at definite points in your presentation; they should arise spontaneously from enthusiasm, conviction and emotion. There are two basic types of gesture: conventional and descriptive.

Figure 4-4. Improper Platform Behavior (continued).

b. Conventional Gestures.

(1) Pointing. The index finger has been used universally to indicate direction and to call attention to objects at which it is pointed. Use it to reinforce an accusation or challenge by pointing directly at the audience or at an imaginary person.

(2) <u>Giving or Receiving</u>. If you were to hand someone a sheet of paper or to hold out your hand to accept one offered to you, the palm of your hand would be facing upward. This same gesture is used to suggest the giving of an idea to the listeners or to request that they give you their support. Sometimes it is combined with the pointing gesture - the idea is held out in one hand while the other hand is used to point to it.

(3) <u>Rejecting</u>. A forward movement of the arm with hands raised and palms turned toward the audience can be used to reinforce such statements as: "This cannot be tolerated", "Hold everything", or "Just a minute".

(4) <u>Clenched Fist</u>. This gesture is generally used with expressions of strong feeling such as anger, power, or determination. The clenched fist may be used to emphasize such statements as: "We must put every ounce of our energy behind this plan". Sometimes the clenched fist is pounded into the other hand or upon the lectern or table.

(5) <u>Dividing</u>. Move the hand from side to side with the palm vertical to indicate the separation of facts or ideas or to divide the audience into imaginary opposing factions.

(6) <u>Restraining</u>. Extend your hand at shoulder height, palm facing outward and downward. This is the gesture to use when saying, for example: "Now, take it easy," "Just a minute," or "We're coming to that."

c. Descriptive Gestures.

(1) Descriptive gestures portray an object or illustrate an action. The speaker describes the size, shape, or movement of an object by imitation. A vigorous punch is shown by striking with the fist; height, by holding the hand at the desired level; speed, by a quick sweep of the arm; complicated or humorous movement, by pantomime as you describe it. The curves of a beautiful girl are described with both hands outlining the lateral figure. Churchill's "V" formed with two fingers was symbolic of victory. Even teenagers graphically describe a "personality" by forming a square with their two hands.

(2) The gestures described above will give you an idea of what might be done with gestures. Your gestures will depend to a large extent on whether your personality is vigorous and dynamic or calm and easy-going. But, regardless of your personality, gestures will add to the effectiveness of your speech if you <u>relax</u> your shoulders, arms, and hands and concentrate on communicating to your audience the meaning and importance of your ideas. When the gesture is natural, it is effective. If the gesture is artificial, posed or strained, it detracts rather than reinforces.

4. <u>FACIAL EXPRESSION</u>. A facial expression is a type of gesture, but it is considered separately here because of its importance. If you are to sway people's minds, inspire them, or even interest them, your face must show what you are feeling and thinking. The most common fault in facial expression is the "dead-pan" or total lack of expression. You can overcome dead-pan by looking over your audience until you find someone who is smiling or has a naturally pleasant face. Smile back at him and you will find yourself warming up to your listeners. Unconsciously, your face will take on meaningful expression. Another common expression difficulty is that of the constantly intense expression, usually manifested by a frown. Overcome this by relaxing all over, then use your intensity only on key ideas. Finally, remember that you are neither a wooden Indian nor a clown - you are a human being; and the more natural and human you appear and act, the more you will influence your listeners. The presentation area is no place for a poker face.

Figure 4-5. Facial Expression Stimulates Listener Reaction.

D. <u>YOUR MIND</u>. THE PROPER SPEAKING ATTITUDE.

   1. <u>GENERAL</u>.

   a. Speech is a means of communicating ideas and thoughts. Sometimes these ideas and thoughts originate in the mind; sometimes they start with outside stimuli which are evaluated and interpreted by the mind and then communicated. In speech you do more than report facts; you also interpret facts and express opinions. Every word has two effects: what it <u>means</u> or denotes and what it <u>implies</u> or connotes. If speaking were merely the production of <u>objective</u> precise language with no emotional implications nor possible misinterpretations, listening would be an unpleasant and boring task indeed! No speaker would communicate anything of his personality. He would not reveal his feelings, his laughter, his tears. His ability to stimulate, convince, persuade, or interest his audience stems primarily from the emotional effect he produces; not necessarily emotion in the sense of joy, sorrow, love and hate, but, more often, emotion in the sense of strong belief, conviction, earnestness, quiet sincerity, or dynamic enthusiasm. Your speech carries with it emotional overtones, how you feel about what you say as well as the emotional impact of your speech upon others. Thus, emotion is the indicator of how a person feels about all that surrounds him - or, more simply, it is the indicator of his <u>attitude</u>. His attitude affects the words he chooses to use. There are four specific indicators of a proper speaking attitude: sincerity, confidence, enthusiasm, and humor.

   b. We cannot think or conjure up images in the mind except in the form of words or objects that have already been classified and typed by a word. We think in terms of words. Every word in the English language has greater or lesser emotional overtones to each listener, depending upon his background. Words like "mother, home, school, hometown" evoke strong emotional reactions in your listeners. Semanticists know the emotional impact of words. The dynamic speaker is aware of word power and a master at using the forceful word at the right time.

   2. <u>SINCERITY</u>.

   a. Sincerity, from the speaker's point of view, is the apparent earnest desire to convince your audience of the truth and value of your ideas. There are two sources of sincerity: (1) Your personal, intense belief in your subject; (2) Your belief in the value of your subject to your listeners. The first of these sources is ideal since all personal belief which is intense enables sincerity to flow forth naturally from every word or gesture of the person. The second source is more rational than emotional. You are convinced that this teaching material is extremely valuable and you present it in an honest and forthright manner. You do not rely on gimmicks or questionable reasoning to make your presentation look good. The word sincerity means without wax (sine/cera) and is derived from the Roman sculptor's trick of covering up chiseling errors by filling them with wax. You won't pad your presentation if you develop belief in the value of the material to students. You'll present it simply and honestly.

b. Assuming that your sincerity originated from one of the sources described, you must show that sincerity! If you appear to believe in what you say, you have taken a major step in convincing your students of the importance of the subject. Sincerity shows in a number of ways: directness of manner, facial expressions, clarity of explanation, the proper combination of humility and authority, effective use of the voice and body to reinforce and emphasize ideas. Regardless of the source of sincerity influencing you, no matter how your personality dictates that you show your sincerity, the important thing is that your students see, hear, and feel your belief in what you say.

3. CONFIDENCE.

a. Confidence is a personal attitude or feeling of assurance, a belief in your ability to perform a task well. In order to be confident, you must have three basic prerequisites: knowledge of your subject, belief in your ability to speak, and the power to control nervousness or "stage fright". Knowledge of subject you obtain through research and study. Belief in your ability comes from rehearsal and experience. Both of these requirements are entirely up to the individual to accomplish in his own way. But the third factor requires further explanation.

b. Control of obvious stage fright increases with proper practice. Steady, regular breathing has a calming effect on the whole body. Relaxing the muscles of the body as much as possible will help to maintain poise and more complete control. Some speakers find that purposeful walking or moving about gets rid of stored-up nervous energy and brings back normal balance and ease. Others conduct a short rehearsal or "warm-up" just before their presentation by running through their introductory remarks. Remember that the problem is control of a natural reaction. If you feel no stage fright or tingling anticipation, you are not taking your job seriously enough to accomplish it effectively.

c. If you stand erect, move purposefully, look your listeners in the eye, and let your ideas flow freely and clearly without stumbling and awkward hesitation, you will appear confident to your audience - and, best of all, you actually will be! Rehearse your introduction thoroughly, for stage fright usually occurs in the first few minutes of a presentation. If you get off to a good beginning, tension will soon disappear.

4. ENTHUSIASM.

a. The word enthusiasm was mentioned previously in describing other aspects of effective speaking. Enthusiasm is the outward manifestation of sincerity and confidence. From the speaker's standpoint, enthusiasm is defined as strong personal excitement or feeling about a cause or a subject.

b. Enthusiasm is not shouting; it is not affected, over-dramatic speech; it is not waving of the arms and leaping about on the platform! Rather, it is the way you show your belief in your subject. You can, indeed you must, be enthusiastic - but remember - each in his own way! Vigorous, dynamic persons will show enthusiasm by brisk, energetic movement, sweeping gestures, a rapid rate of speech, widely-varying inflection, and plenty of vocal force. Instructors of a more subdued nature will move and gesture with less energy, speak in more measured tones, use force only on the key words and ideas, make more use of the pause for effect, and maintain a calm, pleasant, but confident and authoritative manner. Others, probably a majority, will be enthusiastic by combining various characteristics from these two extremes. Don't hide behind a pretense of dignity that takes the form of stiff, deadly monotonous mouthing of words. Loosen up - mentally, emotionally, physically! Stop at nothing, within the bounds of common sense and decency, to persuade, convince, and TEACH! How you show enthusiasm will be governed by your personal characteristics or traits, - but do it you must!

5. <u>HUMOR</u>.

    a. It is entirely possible that an instructor who has the necessary attributes of sincerity, confidence, and enthusiasm may still be lacking in proper attitude to gain student attention, interest, and understanding. A person who lacks a sense of humor will give listeners the impression that he is unreal, inhuman, or very conceited. Humor is the quality that shows you are, after all, just another human being, that you have a warm, lively interest in all that goes on around you. Having a sense of humor does not necessarily imply the ability to tell funny jokes, although there is certainly a place in good instruction for humorous stories of the appropriate type. The thigh-slapping, belly laugh type of humor has its place, but not as a steady diet.

    b. A more effective type of humor is spontaneous classroom humor. Take advantage of unexpected humorous classroom situations which sometimes arise by making a brief comment, a well-placed pause, or by a single open smile. Humor directed at yourself is one of the most effective types. Remember that one of the basic things that make people laugh is the sight of something or someone regarded as important or pompous appearing ridiculous.

    c. In addition to decency, the only rule to follow in using humor is that of good judgement. Take care in directing humor at individual students for classmates may resent this. Be sure your humor is good natured and lightly done. Clean humor is as American as the hot dog and will frequently assist student learning.

CONCLUSION: When class interest is gained, when the demonstrated performance has obtained the responsiveness and receptiveness of the students, only then will you be an effective speaker - only then will you be worthy of the name: Instructor.

YOU HAVE THE TOOLS. Your goal is to make the most of what you have by: (1) Understanding how to speak effectively; (2) Planning improvement; and (3) Practice - all day, every day, wherever you are and to whomever you speak. You have in your hands the means by which you can attain that goal. And while you're practicing, keep this always in mind: It is more difficult to be plain in speech than to be fancy. So be simple rather than artificial, be what you are rather than pretend to be what you are not.

Chapter V

CONTROL OF INTEREST

A. INTRODUCTION.

1. GAIN, MAINTAIN, AND RETAIN STUDENT INTEREST.

    a. There is probably none among us who has not had this experience. You come into a classroom and take your seat. You are one of several students. As you look about you, you notice that the curtains are neatly arranged, the lighting is good, the instructional area is orderly, and there are no distracting influences. The classroom looks right for learning. An instructor begins. He is a young, energetic fellow with a wide variety of experience. He has a pleasing manner and an excellent speaking voice. The instructor begins the lesson with an introduction which stimulates your interest and suggests that something worthwhile is to follow. You recognize that the subject is one that is vital to you. You lean back in your chair, cross your arms and get set to learn. The instructor explains that there are five main points which you must learn during this period in order to be able to use the subject. He explains the first of these five points and discusses several considerations that influence an understanding of this first point. You follow the instructor through point #1 and are with him when he starts to explain point #2. Gradually it happens: The chart appears foggy. The instructor's voice fades away. You are no longer in the classroom. You are busy replaying the third golf shot on the seventh hole yesterday, and wondering whether the trouble was in your grip or in your back swing. Or perhaps you are buckling on your skis for a long run down a snow-covered slope. You are no longer learning. Periodically you snap out of this reverie and try hard to follow the presentation. But again and again you drift down through channels which offer far less resistance than the subject to be learned.

    b. This situation is one which could happen to you. A good lesson plan is not worth much if learning does not take place - and learning will not take place unless you can keep students interested in your presentation. "Interest" in instruction doesn't just happen; it must be carefully gained, maintained and retained in lesson presentation. To help you understand how you will plan to control interest, consider the following:

        (1) How does interest affect learning?

        (2) What are interest factors and how do I use them?

        (3) Is there a span of interest?

        (4) How would a teaching vehicle affect interest?

2. HOW INTEREST AFFECTS LEARNING.

    a. You can make a shooter do many things but you can't make him learn. A student will concentrate, apply, and learn only when he is interested. Some student interest will be self-generated: he may enter your class with a desire to learn this subject, a personal interest in it, and an understanding of why the subject is important to him. To other students you must explain the importance of this subject. Even after student motivation, attention, and concentration has been accomplished, you must give direction, provide usable information and guide the learning in an interesting way. An instructional period is interesting because of what you as instructor say and do and because of what the students do (mentally and physically).

b. At the beginning of the lesson you state the purpose of the lesson and emphasize why it is important for the student to learn this material in order to do his job effectively. The student is motivated--he realizes the need to learn, so he gives his full attention to what the instructor is saying and doing. Now you give direction to this student attention by orienting him on the specific things to be learned.

c. By various methods which involve hearing, seeing, and doing, you present and develop the teaching points which the student must understand and apply to accomplish the learning.

d. Interest is the continued focussing of student attention upon a specific felt need until that need is satisfied. It is the instructor's responsibility to control student interest throughout the lesson. If the student loses interest, he no longer gives his full attention or concentration to the lesson and learning breaks down.

e. How do you maintain student concentration and control interest? By using presentation techniques which appeal to the student and lead to rapid and thorough learning. You must hold student interest by planning for interesting student learning activities--interesting hearing, seeing, and doing. These interesting learning activities are called Interest Factors.

B. INTEREST FACTORS AND HOW TO USE THEM.

1. INTEREST FACTORS ARE NOT GIMMICKS. Interest factors are the presentation methods and techniques used to gain, maintain or increase student interest in instruction. These are not gimmicks thrown into instruction merely to startle students or gain momentary attention. Interest factors are ways by which the instructor communicates his teaching points and supporting ideas - they must assist the presentation. To be effective, interest factors must help to maintain attention and concentration on the material being presented. Analyze different interest factors according to the physical processes by which they lead to learning: hearing, seeing and doing.

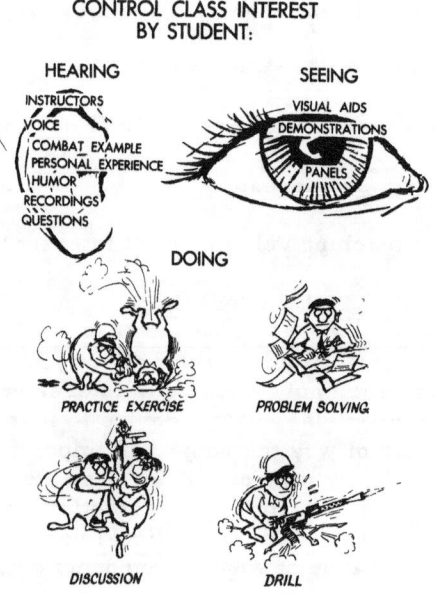

Figure 5-1. Control Class Interest.

2. <u>HEARING</u>. The Introduction to this chapter described how the instructor talked for the entire period. He failed to control interest because continuous verbal explanation is not interesting. Explanations can be and must be brought to life and made interesting by combat (or historical) examples, personal experience, humor, recordings, and questions.

    a. <u>Instructor's Voice</u>. Your voice can be an excellent interest factor or a boring feature of your instruction. Radio and television announcers, commentators and talent make their living through skillful voice techniques. They captivate listener interest by varied inflection, significant pauses, clear articulation and pronunciation, and skillful variations in rate and volume. A good portion of your instructional success will depend on your voice delivery. You may possess all the knowledge in the world but unless you communicate this knowledge in an interesting, dynamic manner you are an ineffective instructor. The mere sound of your voice may gain interest initially but how long will your voice hold interest? If you speak monotonously or mumble or swallow your words, you'll lose your student audience within a few minutes. No student will struggle to understand you no matter how important your lesson may be.

    b. <u>Personal Experiences.</u> Personal experiences lend the weight of authority to the points you are making. They may be taken from your own experience or may be quotations from other sources. To be interesting and effective for learning, personal experiences must be based on valid authority. Personal experience must be based on impressive experience or observation over a long period. You must exercise care in giving personal opinion to be sure that your experience in the field which you are discussing is sufficient to make such opinion valid. This limitation does not apply to factual experience, particularly that encountered in combat. Factual personal experiences are just that: facts that you have experienced.

    c. <u>Humor</u>.

        (1) Humorous stories and illustrations relax students and help them to remember the points emphasized by humor. Marksmanship instruction deals with deadly serious ideas. Unrelieved tension leads to the mental ward. Use humor to relax student tension momentarily. It will assist students to maintain interest in the subject. Don't restrict humor to joke telling alone, include skits, demonstrations, cartoon slides, and training films. Capitalize on spontaneous humor when it arises in a teaching situation, as from a funny, unexpected reply during a discussion. The instructor who displays a keen sense of humor has won most of the battle for control of student interest.

        (2) Humor must be clean and must be related to the instruction. The group that gathers behind the club house to exchange dirty stories is a volunteer group. Members may leave at any time when they think they've had enough. Students in class are a captive audience; they do not have the option of leaving when foul or obscene jokes are told. For this reason, if for no other, the instructor should respect the rights of all his students and should use only humor which is clean. He should not ridicule any race, religion or nationality. To do so alienates those students concerned and others sympathetic to those offended. From another standpoint, the instructor must set a high standard in his humor to retain the respect of students. Telling dirty stories is a cheap appeal for popularity with the coarsest members of the class.

        (3) Humor will not be used solely for entertainment. Humor must teach or it may distract, destroying the student's concentration on the subject. Tie in humor very closely with a specific idea in your lesson so that when the student recalls the humor, he also remembers the point you made.

    d. <u>Recordings</u>. Recordings add realism and variety to your presentations; they bring into the classroom voices and sounds not otherwise available. Use recordings alone, or in conjunction with skits, demonstrations, or projections. Bring into the training area the voices of such leaders as MacArthur, Eisenhower and Churchill. Record a conversation and thus avoid the need for many assistants to be physically present to act out such a conversation. Recordings are easy to prepare and use.

      c. <u>Questions</u>. Thought-provoking questions force students to concentrate on the instruction to prepare an appropriate answer if called upon. There are several types of questions you can use, depending on your purpose. These are discussed in Chapter 6. Generally speaking, the best questions are those which stimulate student analysis of a situation and require him to reason through to a second solution. The effect of a question is to alert the student to the possibility that he may be called upon. He concentrates on the material being presented so as not to be caught unaware and perhaps embarrassed by not knowing the answer to a question. Therefore, a good question should be asked in such a way as to give most students ample time to formulate a reply. If you begin your questions with "What," "Why," "How," etc., you gain the students' attention and they are alert to the critical part of your question. Don't put these pronouns at the end of the question.

    3. <u>SEEING</u>. "A picture is worth a thousand words." This is an old but true saying. Some students are able to grasp ideas that are developed by verbal explanation and discussion alone. But for most students, real understanding comes only when they can "see" the ideas being discussed. In some learning situations, it is fairly obvious to the instructor that he must "show" something for full student learning. In many cases, however, this need to "show" is not immediately apparent. Use imagination and ingenuity to provide ways for students to learn by seeing as well as by hearing. Otherwise, students will lose track of the idea being developed. If the idea is explained in conjunction with a visual demonstration of the point, student concentration will be reinforced by the "seeing" activity and better learning will result because students are able to follow the idea visually. Ways of holding interest by "seeing" are visual aids, demonstrations, skits and panels.

      a. <u>Visual Aids</u>. Visual aids may be interest factors, but not all visual aids are necessarily interest factors. A well selected aid, properly used, will be interesting because it assists student understanding. The types of aid and their uses are covered in detail in Chapter 7, "Training Aids." As far as interest is concerned, visual aids must focus student concentration on the points to be learned. A simple, clear aid will focus student thought on that idea. A complicated aid will only confuse the student with too many ideas and distract his attention from the point you want to emphasize. Color in an aid may be used to direct student attention to a particular point. An aid must be large enough to be seen by all; if it is too small to be seen clearly, student interest is lost. The aid must be introduced at precisely the correct, logical point in the learning situation, and must accurately portray the idea being developed by explanation.

      b. <u>Demonstrations, Skits, Panels</u>. These are all forms of the same thing: a dramatic representation of an idea. Demonstrations show how something is done. Skits may portray a procedure or operations, usually with a humorous slant; they are particularly appropriate to demonstrate personal interrelationships in team coaching. A panel is a group discussion, usually among various instructors, although sometimes it includes students. Panels will maintain student interest if the discussion is closely controlled so that it does not wander from the subject. All three of these forms of presentation must be well planned and rehearsed. They must be designed primarily to teach, not merely to entertain. Use humor to increase interest in demonstrations, skits, and panels but don't overuse it.

    4. <u>DOING</u>: DISCUSSION AND PRACTICAL WORK

      a. <u>Students Learn by Doing</u>. Interest is high when students are doing. Doing activities include discussion and all types of practical exercise. Doing may be mental or physical.

b. _Discussion._ Student discussion is a form of mental doing. Discussion, when animated by thought-provoking questions, is an excellent interest factor. When students become involved in instruction to the point of freely expressing their ideas in your discussion, class interest is high. To learn, the student must become involved in the subject intellectually and emotionally. However, you must control and direct the discussion so that the major points of the lesson are developed. A wandering discussion may be interesting, but students will learn little unless you channel the discussion and make periodic summaries. This is your job as monitor of student discussion.

c. _Practical Exercises._

(1) Most marksmanship instruction teaches practical knowledge and skills - the ability to perform specific tasks. Practical work provides the best means to teach these skills and knowledge. Too often, the instructor spends too much time on lengthy explanation, and allocates too little time to student doing. Some orientation and direction must preface learning to do things. But the sooner the student begins doing, the higher his interest will be. You may use practical exercises with any subject.

(2) To be interesting, practical work must give all students a specific task, and the time allotted for the task should be just enough to complete the job. With too much time, students lose interest; with too little time, students become frustrated.

5. CONTROLLING INTEREST vs ENTERTAINING.

a. Interest factors are interesting ways or techniques of presenting the ideas of a lesson. This chapter discusses only a few; there are many other interesting methods, limited only by imagination. Interest factors used with imagination together with good instructional techniques help students learn. Some years ago, an instructor at USAIS was widely acclaimed as the most interesting instructor in the Infantry. But few students could tell you what he taught. He had entertained extremely well but students did not learn much about his subject. He overdid interest to the detriment of learning. Don't overuse interest factors to the point that students actually lose interest and become bored. They tire of too many jokes, extended periods of practical work, too much discussion, too many training aids. Once a point is learned, students want to move on to another. If a point has been well learned, through explanation, examples and discussion, further explanation of the same idea will bore the student.

b. Overuse of one type of interest factor will lessen its appeal as the lesson progresses. We know that one interest factor holds high student interest only for a short time - then interest decreases unless something new picks up again.

C. INTEREST SPAN. Interest span is the length of time that a student can give undivided attention (interest) to instruction.

1. LENGTH OF INTEREST SPAN. The interest span depends upon many factors. One very important factor is the instructor's presentation techniques. Many additional factors influence the interest span. Some subjects interest students more than other subjects. Personal physical comfort of students affects their ability to concentrate; weather conditions influence the span. Students concentrate longer and better in the early hours of the day when their minds and bodies are fresh and rested. Other factors being equal, the interest span shortens when the instructor is talking, and the span lengthens when students "do." The instructor with a knowledge of human relations quickly senses a drop in class interest and promptly modifies his actions to regain student interest.

2. BRIDGING THE INTEREST SPAN. Instruction cannot consist solely of student performance. Before the student may perform effectively, he must learn what and how to perform.

This type of instruction invovles instructor presentation by lecture, conference, and demonstration. In this instruction, too often the instructor lectures at great length without concern for student interest. As a result, students lose interest after a certain amount of lecturing - they reach the end of their interest span for that particular part of the instruction. Carefully analyze the materials for probable dull spots in your lesson and plan to maintain interest by logical integration of interest factors to bridge the interest spans. For instance, you may begin with an attention-gaining introduction, then find that the next fifteen minutes are devoted to lengthy explanations. You know that, at best, you can hold interest for about five to ten minutes while lecturing. So you must plan to use an example, an aid, humor, or other interest factors early in this period to keep student interest high and to maintain it throughout this fifteen minutes. Throughout the lesson interest tends to drop after students learn a point. The instructor must re-motivate students periodically with additional interest factors to reduce or eliminate places in his instruction where interest is low.

D. TEACHING VEHICLES.

1. AN INSTRUCTIONAL DEVICE. A teaching vehicle is an instructional device which binds the teaching points logically around a central theme or story and helps to give interest, movement and continuity to a presentation. During a period of instruction, the instructor uses many examples and illustrations, humor, quotations, and other interest factors. Students tend to become confused if these factors are unrelated. However, the instructor can make his instruction more logical and meaningful by relating interest factors and supporting ideas to a single, continuing situation. This situation may be developed by the use of actual equipment, practical exercises, combat and historical examples, or by creating a fictional character or situation.

2. TYPES OF TEACHING VEHICLE.

a. Actual Equipment. Students follow easily the use of actual equipment, where the natural relationship of various parts and their operation are physically apparent. The discussion of points should follow some logical pattern in the construction and operation of the equipment. One interesting method is to use the background and history of the equipment to show its development from its earliest form. For example, one instructor taught the steering mechanism of a 2 1/2-ton truck by beginning with the earliest form of steering used in horse-drawn wagons. From this point, he discussed each successive advance in the engineering of steering mechanism until he had developed a complete understanding of present day equipment. He used historical development as the teaching vehicle.

b. Continuing Situations. Don't restrict use of continuing situations to tactical subjects. Continuing situations help to teach any subject. They may take the form of handouts, skits, or just explanation. Using a continuous situation to present problems for

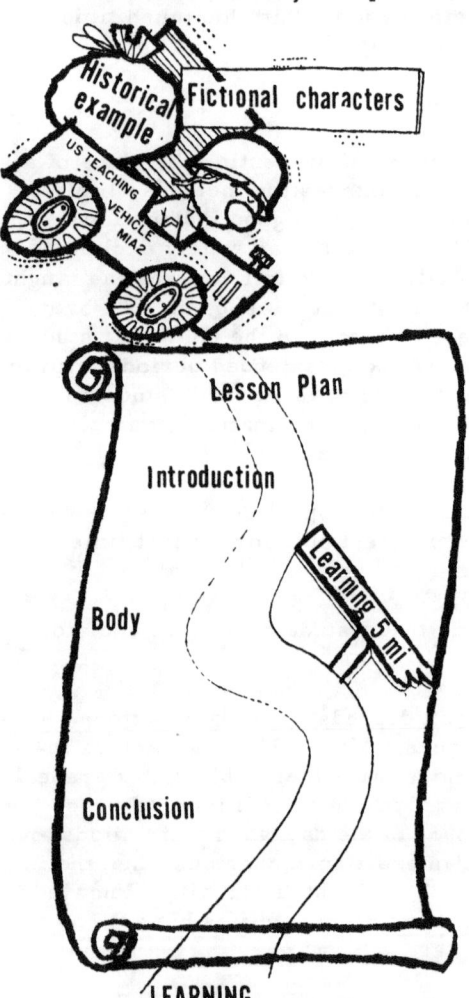

Figure 5-2. Teaching Vehicles Unify Instruction.

students to solve and discuss not only creates interest, but makes the subject matter more realistic and learning more effective.

c. <u>Historical Examples</u>. Historical examples make good teaching vehicles. They may form the basis for practical, or help to explain different teaching points in a lesson. Details of what actually happened in the example, pertinent quotations and word pictures add interest and insight.

d. <u>Fictional Characters and Situations</u>. An instructor with imagination may create a fictional situation and character to fit his problem. Place your fictional character in continuing situations to develop teaching points in a realistic and interesting manner. Few people would read an involved analysis of the psychological impact of the Civil War. Yet millions of people paid to read the story of Scarlett O'Hara and Rhett Butler, in "Gone with the Wind." Margaret Mitchel, by creating a fictional family of this period, became world famous.

3. <u>VALUE OF A TEACHING VEHICLE</u>. The teaching vehicle is an excellent method for "continuing the situation" to show the logical relationship or development of your teaching points. It helps to tie together interest factors and to make logical transitions from point to point in the lesson. It smooths the humps and valleys of the interest span.

E. <u>APPLYING CONTROL OF INTEREST</u>.

1. <u>CONTROL OF INTEREST FOLLOWS LESSON OUTLINING</u>. We have discussed interest, interest factors, interest span, and teaching vehicle separately. How do you apply them to instruction? You don't just sit down and write an interesting lesson. You plan for interest after drafting a good sentence outline with your ideas completely expressed and logically organized. With this outline drafted you are ready to apply techniques to control interest.

2. <u>SELECT A TEACHING VEHICLE</u>. First, consider the use of an overall teaching vehicle appropriate for the whole lesson. This will require imagination and ingenuity. The situation may be true or imaginary, but it should be realistic, one which students may later face. Teaching vehicles lend themselves well to the use of humor, skits, and demonstrations which relate to the basic story. This vehicle will serve as a logical framework upon which to develop the ideas of the lesson and supply a reference point for smooth transitions.

3. <u>ANALYZE THE INTEREST SPANS</u>. Next, analyze your outline to determine places where the interest span is short, where interest will tend to decrease. These portions of the outline will require interest factors. Interest is usually lowest when you are lecturing for extended periods of time with no interest factors used.

4. <u>INTEGRATE INTEREST FACTORS</u>. For these areas where interest is low, plan to integrate interest factors to bridge the interest span. The interest factors must relate to the material, and should, if possible, tie into the teaching vehicle. By using interest factors, you will raise student interest at these points, and this enables you to emphasize your main ideas while interest is high. You try to keep interest high throughout the lesson, but you must insure it is high at those times when you stress the main points of your lesson.

F. <u>A LESSON WITH CONTROL OF INTEREST</u>.

1. <u>HOW TO BRIDGE THE DULL SPOTS</u>. In the introduction to this chapter, we described a situation where the instructor has a good outline of ideas but failed to control student interest. A graph of what happened appears in Figure 5-3. If he used appropriate interest factors, a graph of interest control during his lesson appears as shown in Figure 5-4.

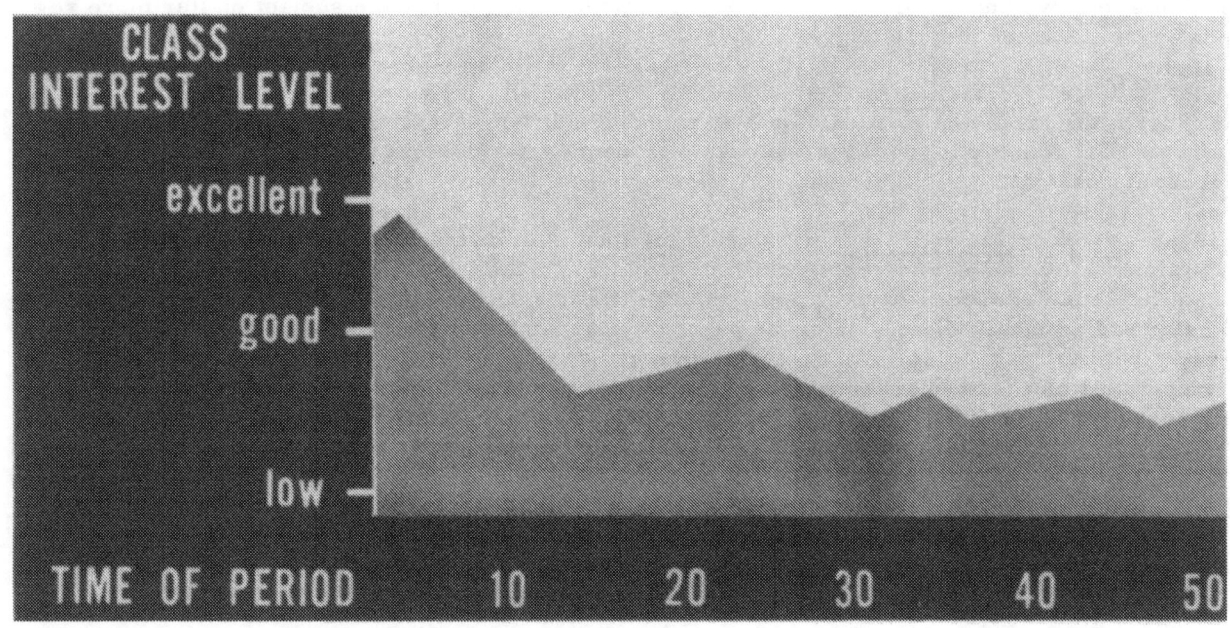

Figure 5-3. Graph of Lesson with Poor Control of Interest.

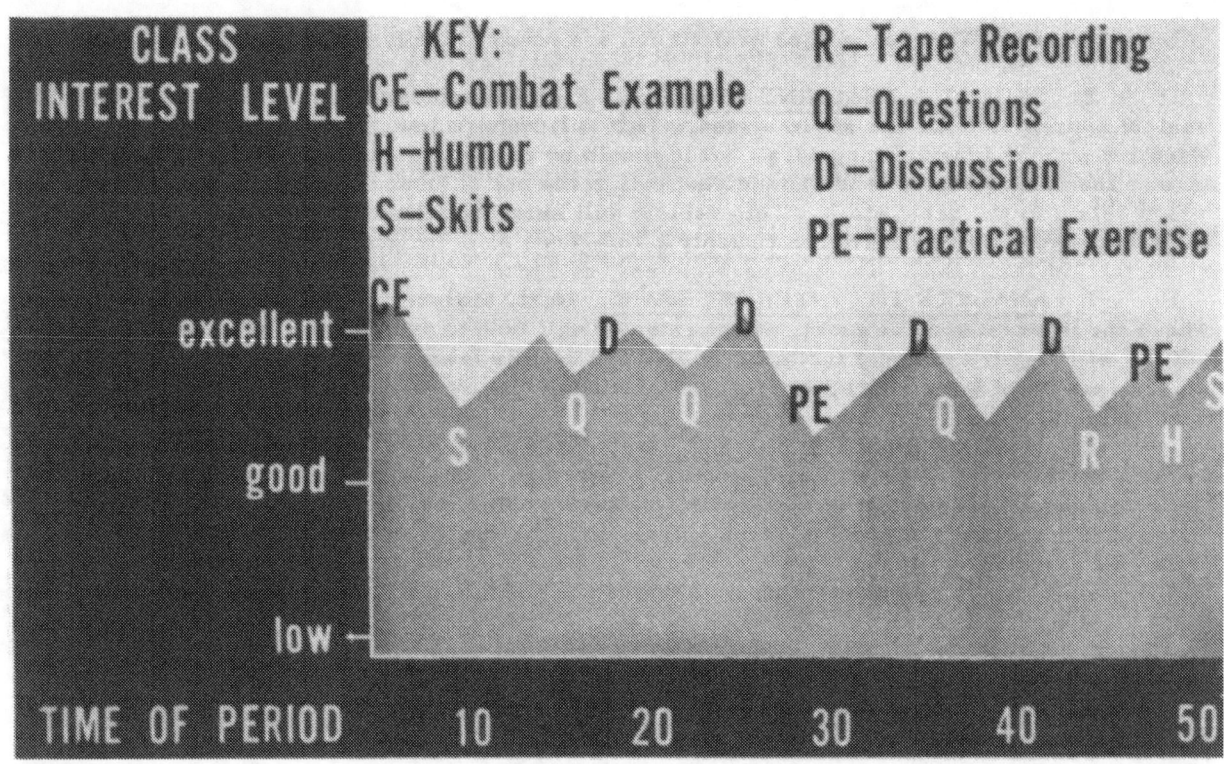

Figure 5-4. Graph of Lesson with Good Control of Interest.

2. HOW THE NEW LESSON DEVELOPS.

    a. In Figure 5-4, he used an example in his introduction. Then he began to explain the first point. While he was talking, interest went down. So he used a humorous skit to demonstrate his point. Interest rose, and he capitalized on it by asking questions and stimulating discussion. Interest dropped slowly as the discussion progressed. At this point, he involved the students in a practical exercise. Interest was high again, but fell off as students solved the requirement. He again picked up interest with student discussion of solutions. He concluded the discussion with a tape recording of a sound solution to the exercise.

    b. He issued another practical work exercise requirement followed by a short discussion of solutions. At this point, the student interest span was shorter, since the class was almost over. So he used a humorous story to raise interest high, and concluded with a skit which re-emphasized the main points of the lesson.

    c. As you can see in the graph, interest remained above the "good" level throughout. He had learned that students can concentrate on instruction only when he assisted them by making his presentation interesting. By picking up student interest at critical points in his lesson, he assured continued student concentration and its logical consequence - more effective student learning.

Chapter VI

MANAGEMENT OF INSTRUCTION

A. GENERAL

1. GOOD MANAGEMENT INSURES ENJOYABLE TEACHING:

    a. Instructional management and classroom administration may seem to be only indirectly related to instructional effectiveness; nevertheless, these administrative aspects of your work are of great importance. They set the tone for the entire instructional job and can determine, to a large extent, whether your students learn. Attention to the conditions of instructional environment, to the needs of students, and to the proper supervision of their activities will lubricate the machinery of instruction to reduce classroom friction or confusion. The better you systematize the routine tasks associated with teaching and the more you eliminate physical, mental, and personal distractions, the more time you will have to concentrate on the most enjoyable part of your work - teaching.

    b. Although most of your instructional time will be spent in lesson preparation, the best prepared lesson will fail if you do not provide proper classroom and instructional management. During the actual presentation, many situations will occur for which you can not possibly plan. It is essential, then, that you be aware of the common types of difficult situations and be flexible enough to cope with them. Section B will discuss the problem of questions used during a conference type presentation. Sections C, D, and E will be concerned with the instructor's control of the class through subsummaries, presentation of solutions, transitional techniques, and handling of difficult conference situations. Section G will consider the need for good student/instructor relations in teaching, and finally, Section H will outline some suggestions for better classroom administration.

B. QUESTIONS

1. GOOD QUESTIONS ARE THE HEART OF THE CONFERENCE.

    a. Good questions are the heart of the conference since they are used to stimulate student thinking, to check understanding and retention, and to hold attention. The primary purpose of questions during a conference is to guide thinking or lead to understanding through discussion; secondarily, they create interest and hold attention.

    b. Questions should have a specific purpose, be clearly worded, and thought provoking. The specific purpose of most conference questions is to guide student thinking along pre-determined lines which lead to sound reasoning and understanding. Phrase the questions in simple form so students will have no difficulty in recognizing what is required. Simple and well worded questions will not just happen the first time you write them out (and you __must__ write them out). Phrase them and rephrase them until the questions ask what you want. Like good test questions, simple, clear conference questions are the result of much instructor effort. They should stimulate thoughtful reasoning by the student - emphasizing the "how" and "why" - rather than asking for rote memory or recall.

    c. Try out the questions on your colleagues. See how they answer them. If your fellow instructors are confused and request further clarification your students will undoubtedly be confused too.

    d. There are four types of planned questions which you will use during a conference lead-off questions, follow-up, check-up, and rhetorical.

2. LEAD-OFF QUESTION.

   a. The first of these conference questions is the lead-off question. This is the question which initiates discussion on a selected teaching point. The lead-off question is general in nature to guide thinking along broad lines which may encompass one or more of your planned teaching points, or the whole subject. It is used to focus attention on "the big picture" as a preliminary to more specific discussion of particular points. Therefore it is a good technique to address the lead-off question to several students so as to generate broad introductory discussion of the subject. Your lead-off question must be carefully planned so that student interest is immediately aroused and a variety of responses is anticipated.

   b. The lead-off question should be provocative and should be so planned and stated that "yes" or "no" answers or short obvious answers are not possible. Normally, questions which begin with "why", "how", and "under what circumstances" are questions which will generate thinking and discussion. Should it be necessary in planning a lead-off question to have one which will require a "yes" or "no" answer, follow up immediately with a "why" or "why not". Lead-off questions, like teaching points, will be preplanned and inserted in the lesson outline in the place where they normally will be used. A good question is like a good transition. Once it has been proved effective, it seldom needs to be changed or revised. Sometimes conferences prove ineffective solely because the lead-off questions were not thoroughly planned.

   c. An example of an acceptable lead-off question might be: "Why is it necessary to make sight changes"? This type of lead-off question is thought-provoking and would normally generate discussion.

3. FOLLOW-UP QUESTION.

   a. Another type of question used in conducting a conference is the follow-up question. Follow-up questions are questions asked as a result of student responses to lead-off questions. The instructor uses follow-up questions to guide the discussion toward an understanding of the teaching point by a consideration of the more detailed or specific aspects of that particular point.

   b. Certainly, you cannot know in advance the actual responses which you will receive from students. Thus you cannot plan the exact wording of your follow-up questions in advance. However, you should consider possible student answers to lead-off questions so that you can have some advance idea of the type of follow-up question which will be required.

   c. Flexibility is required in order to ask effective follow-up questions. Thorough research is needed in order to become aware of the usual student answers. Follow-up questions will not be an integral part of the lesson outline but reference to possible student answers and suggested follow-up questions will be placed in the outline in the form of notes.

   d. An example of a follow-up question follows: If a student answered a lead-off question with a short answer which did not indicate the process which he considered in stating his answer you might follow up with, "Explain your reasoning in arriving at this conclusion". Or you might ask, "Under what circumstances would your answer not be appropriate?" Another type of follow-up question which generates discussion is one which asks another student to comment upon a student answer.

   e. In conducting a conference you must insure that by student discussion or by your explanation every area necessary to student understanding of the teaching point is developed. By combining good follow-up questions and instructor explanation, you can accomplish this understanding.

4. **CHECK-UP QUESTION.**

   a. A third type of question which is asked during a conference is the check-up question. Check-up questions are factual type questions which check understanding, retention, and recall of points previously studied or developed. They are summary-type questions and are equally useful during a lecture or conference.

   b. Use check-up questions to evaluate how thoroughly students understand a teaching area which has been developed. Check-up questions assist you in determining how much more time should be spent in the development of the teaching point and how much reteaching may be necessary later on in the conference. Some check-up questions should be planned in advance and others should be asked whenever the instructor senses a need to check student understanding. If you suspect that the class may be confused or if student answers are incorrect, ask a check-up question to evaluate the present state of class learning. Planned check-up questions will be inserted in the outline at the place where they normally may be asked. Whether you ask them depends on how the class is progressing.

5. **RHETORICAL QUESTION.**

   a. The fourth type of question which is asked during the instructional conference can be used with virtually any method of instruction. This is the rhetorical questions, a question which is asked and answered by the instructor. One effective use of a rhetorical question is to assist in gaining the student's attention during the lesson introduction. Rhetorical questions are used frequently as a transition or to introduce new areas of discussion.

   b. Rhetorical questions are planned and placed in the lesson outline. The technique of asking a rhetorical question is similar to that used in asking any type of question: ask the question, pause, but then answer it yourself. Don't overuse rhetorical questions or you will lessen their effectiveness.

6. **TECHNIQUE OF ASKING A QUESTION.**

   a. The A. P. C. Technique of questioning is the approved technique. "Ask the question, Pause, and then Call on the student to answer." This procedure is the only acceptable procedure to use in asking lead-off questions. The pause is necessary in order to provide time for the students to formulate their response.

   b. This questioning technique is not an inflexible procedure. Once a discussion has been generated, interest in the discussion may lessen if the instructor pauses between each follow-up question. At such a time you should vary the technique, insuring only that the question is understood and that the person who is designated to answer the question knows that he is required to answer it.

   c. Initially you should call on students to answer questions after consulting the class roster. A technique for insuring that your pause is timely enough is to ask the question, then go to the roster, select the man, check off his name, and call upon the student. In using a class roster it is wise to skip from page to page and not go through the roster systematically one page at a time. After a class has been with you for a short time, each student learns the page of the roster which contains his name and thus he is aware when he may or may not be called upon. This may detract from the effectiveness of the discussion. When asking follow-up questions, it is perfectly permissible to accept volunteer answers. Care should be taken to insure that a small group of volunteers do not monopolize the student discussion. You should use the class roster to insure that this does not occur.

d. As a general rule, don't repeat questions for the benefit of inattentive students, as this will lessen the overall attention holding effect of the questioning. If students know you will repeat everything, they pay less heed to what you are asking initially.

e. When asking questions, demonstrate a sincere and enthusiastic manner. Indicate a real desire to hear the student's viewpoint. Your attitude will affect the answer you will receive and influence the amount of student participation. Interest and enthusiasm are contagious.

2. HANDLING OF STUDENT QUESTIONS.

a. Discussion thus far has concerned the questions which the instructor will ask during a period of instruction. There has been no discussion of the type of question that a student may ask. If you have properly motivated the class, interest in the subject will be high and you certainly can expect student questions from time to time as the discussion develops. Your response to student questions will play an important part in determining the overall effectiveness of your presentation.

b. When a student asks a question that seems unrelated or if he seems to challenge or question something you have said, your first reaction may be, "Oh, a sharpshooter - well I'll cut him down to size". Don't give in to the temptation to be sarcastic. The chances are that since you are somewhat tense while presenting instruction, you may misinterpret an innocent question as an attempt to "shoot you down". Listen to the student carefully and give him the benefit of the doubt. By his second or third question you can decide whether he meant to be antagonistic. Even if you discover that the student is baiting you, don't lose your temper for then everything is lost. Tell him to see you at the break and you'll discuss it further. Your duty to the class is to avoid personal bickering.

c. If a discussion is interesting and student interest is obviously high, it is a good technique to ask another member of the class to answer a question posed by the student. This is especially true if students raise hands immediately following a student question. They probably are eager to respond to the student's question and this is one way to generate lively discussion. If the question is one that you feel the average student may not be able to answer adequately, answer the question yourself. Your guide should be to adopt the course of action which maintains high student interest in the discussion.

d. Even though you have properly prepared yourself and have carefully researched each area which will be developed, from time to time students will ask questions which you cannot answer without further research. When this happens, the accepted course of action is to inform the students that you will find the answer and let them know. After you have committed yourself to do this, you must insure that the students get the information.

8. REACTION TO STUDENT ANSWERS.

a. Remember that when a student is called upon, he should rise, face the majority of the class, state his name and speak loud enough for all to hear. Use tact in this regard. If the summer fans are on, it may be worthwhile to use your mike power to summarize or repeat his words. Indicate when you are through with a student so he may be seated; don't let him stand while you discourse.

b. Select from the varied student responses those ideas or facts which you can draw on for further questions and discussion. Recognize a good answer by complimenting the student; this will encourage further participation. Remember that students are human beings who like and need praise. Don't ridicule a poor answer; try to pick out some good point in the reply which you may use to further the class discussion and understanding.

c. While the student is answering, don't turn your back or look at notes. Show an alert positive interest in what he is saying because you must evaluate this response as he is talking and plan your next move as soon as he is finished. Watch that the response and follow-up-questions do not generate into a private conversation between you and one student.

C. SUB-SUMMARIES

1. SUB-SUMMARIES AID STUDENT UNDERSTANDING.

a. Sub-summaries are extremely important when using the conference method of teaching. After active student participation and discussion, the students frequently need someone to wrap up what has been learned as a result of this discussion and relate it to the present teaching point or to the overall lesson objective. In the student discussion some comments and opinions contained key ideas and others were interesting but not too important or relevant. This is a good time to step in and present a summary of this discussion, indicating which ideas are worthy of retention and how they support the teaching point discussed or the lesson objective. Since this is not the final summary (which is part of the lesson conclusion), it is called a sub-summary.

b. A sub-summary is one way that the instructor controls discussion and prevents the class from wandering afield or overdiscussing a certain teaching point. As you sub-summarize you have class control and can use the sub-summary to transition into your next teaching point. After a discussion wherein varying or opposed viewpoints have been expressed, students look to you to comment and, if appropriate, to present the approved position.

D. TRANSITIONS

1. CLASS CONTROL BY TRANSITION. In each problem you develop the teaching material in the way that makes sense to you and, you hope, makes sense to your students. You arrange certain teaching points or supporting materials before others because the lesson is more understandable that way. However, you cannot assume that every class member will see this logical pattern instinctively. This is where well planned transitions are valuable. They are inserted to help bridge over the train of thought from idea #1 to idea #2. They are most essential and must be skillfully used when teaching by the conference method. By good transitions you control or guide the student's thinking along the proper channels and assist him in understanding the logical relationship between key ideas. An understanding of this relationship is essential for a student learning instructional material. Smooth transitions are a sign of professional skill. New instructors frequently fail to make smooth transitions either because they haven't planned them carefully or because they themselves aren't too aware of the relationship of the ideas.

2. TRANSITION BY SUB-SUMMARY. Section C explained how to use a sub-summary to pin down the key ideas which arose from the student discussion and then move on to the next point. A sub-summary is probably the most common way of transitioning into the next point.

3. TRANSITION BY TEACHING VEHICLE. In Chapter 5 the teaching vehicle was defined and discussed. By referring to the teaching vehicle you lend unity to the lesson and indicate how the next teaching point should follow logically at this time. You show the relative position of this teaching point or student performance objective within the context of the whole lesson.

4. TRANSITION BY A SLIDE. Some instructors place the teaching points in a logical sequence on a slide. They use the slide in the introduction to show the scope of the period, then refer to it to transition to each new teaching point. They use the slide again during the conclusion to summarize what has been discussed or learned during the period.

5. TRANSITION BY RHETORICAL QUESTION. Another method of transition is to use a rhetorical question. Instructors frequently transition in this way: "Now that we have discussed A and B, what is the next logical point to consider?" (pause) "C". Then they explain why C is logically next.

6. TRANSITION BY NATURAL PATTERN OR SEQUENCE. Certain instructional material lends itself to a definite natural pattern of development. For example, in discussing a motor vehicle it would be natural to consider various systems in order, such as ignition, fuel, brakes. In a problem situation you would consider events in a chronological pattern of what happens first, then what happens. These subjects contain natural transitions. To aid understanding all you have to do is indicate that you are finished with that phase and are now going into a new phase. Again, here the chart or training aid may help as a transitional device. However, a word of caution, don't simply display a new chart without making some introduction or transition into it. Students should be adequately introduced to new material whether it is visual or verbal material.

E. RESOLVING DIFFICULT CONFERENCE SITUATIONS

1. STUDENTS WON'T DISCUSS.

    a. Occasionally you will encounter a class wherein the students simply reply "I don't know". Try to diagnose the situation on the spot. This is where your human relations skill as an instructor is important. Perhaps you might relax them with a humorous anecdote. Adopt a friendly sympathetic attitude. Rephrase the questions so as to give students additional time to think. Ask them what is their opinion on the subject being discussed or what action they would take rather than what action should be taken. Anyone can express a personal opinion on the subject but not everyone is confident that he has the correct answer. Then get another student to comment or tell what action he would take.

    b. Perhaps you have not yet sufficiently clarified the issue or point. Amplify your explanation and then try the lead-off question again.

    c. Give an illustration which is absolutely contrary to doctrine. This should arouse class discussion because some students should see the error of such reasoning. Then you can lead the class around to the correct procedures.

2. THE STUDENT ANSWERS WITH THE COMPLETE SOLUTION. Sometimes a student will respond with a complete solution and all the supporting material. He apparently has stolen all your thunder. In this case, thank him, and praise the answer. Then you may decide to resummarize what he has stated, adding some additional illustrations or examples. Your summary will be your words but will reinforce the same key ideas. Again you may decide to jot down his statements on the blackboard and discuss each in more detail. Stay flexible and keep your sense of humor. Don't get panicky. A statement of the teaching points by a student does not mean that they have been taught to the class. Another possible tack is to propose alternate actions contrary to the solution and ask students why they would not follow these alternates.

3. **THE STUDENT TALKS TOO MUCH.** Sometimes a student is enamoured with his own voice and rambles along endlessly. You cannot encourage or permit lengthy solo dissertations since the conference is designed to get opinions and discussion from many class members. Hear him out for a while and if it becomes apparent that he doesn't intend to conclude, interrupt him and ask him to condense his comments or summarize. If he still continues, thank him in a polite, definite and forceful way and call upon someone else to comment.

4. **THE DISCUSSION TURNS INTO ARGUMENT.** Sometimes a lively discussion turns into an argument either between students or with the instructor. Some emotional comments may be worthwhile in a sincere, lively discussion. However, don't let comments degenerate into personal attacks. Step in and rechannel the discussion. Either call upon another student for his opinion, settle the issue by a comment reflecting the approved position, or ease the situation with a light remark or a summary.

5. **STUDENTS WHO OPPOSE EVERYTHING.** A student may try to object to everything you say. Rather than argue with him directly, check to see if any hands go up. The chances are that they are from students who wish to object or answer the obnoxious student. Whenever you get students to answer and reflect the approved policy it eases the burden of explaining everything yourself or of putting yourself on the defensive. If the student still persists, remember that (as principal instructor) you are in complete charge of that class period. Ask him to see you after class for further discussion. Remind him that he is taking up class time. If this extremity is necessary, ask him to leave.

6. **THE DISCUSSION GENERATES TOO LATE IN THE PERIOD.**

    a. A class may get into a lively discussion just before the period is scheduled to end. If you have this same class for the next consecutive period there is no problem. You can let the discussion continue for five or ten minutes more. Then you can give them their break a bit later but you will have to make up time lost in your next period. However, do not shorten the class break time.

    b. If the discussion starts late in the period and you have only enough time for a rapid condensed conclusion, you must break in and begin your conclusion. That is where a flexible conclusion is useful. Perhaps you planned five or ten minutes to use a check-up question type of summary. If the discussion was lively, switch to a summary slide and hit only the high points. As a good instructor you must be ready for either eventuality.

F. HUMAN RELATIONS IN TEACHING

1. **NEED FOR CLASSROOM AND INSTRUCTIONAL MANAGEMENT.** Classroom and instructional management are necessary factors in good instruction. The instructor must be concerned with the comfort of his student, the effective and economical use of materials, the timely use of equipment, and the full utilization of all available training facilities. Furthermore, the principles of instructional management and administration are of primary importance in preventing difficulties and in solving problems that develop in a learning situation. As an instructor, you will be called upon to exercise judgment and work out solutions which do justice to everyone concerned. The principles and procedures you adopt in handling this part of your job will go a long way toward making you a better instructor.

2. **THE INSTRUCTOR-STUDENT RELATIONSHIP.**

    a. An instructor must establish a receptive, cooperative, working relationship if his instruction is to be effective. If learning is to be student-centered, the student must be treated as an adult and progressively given more and more responsibility; otherwise, he will gain little or no confidence in performing the task.

b. In addition to assisting the student in gaining a sense of responsibility and confidence, the instructor must develop the sense of belongingness so important for effective student learning and participation. You must take a personal interest in the student and continually strive to raise his personal standard as well as his performance standard.

c. Students respond to and tend to reflect your attitude. If you are natural, helpful, and enthusiastic, your students will be friendly, eager, and ready to learn. Learn their names, talk with them personally whenever you have the chance. Show that you are trying to help them. You are working with men, not machines. Remember, you desire students to be good shooters through the exercise of their own initiative rather than through a conformity which is forced upon them by a superior. If a student is to respect rather than to fear or resent authority, the instructor must be fair, firm, and friendly. Some suggestions which will aid you in acquiring these very important human relations qualities follow.

3. SOME SUGGESTIONS FOR MAINTAINING GOOD RELATIONSHIPS.

a. Show no partiality or favoritism: Nothing destroys student readiness or receptiveness as partiality. Never reprimand a class for the wrong doings of a few.

b. Never try to bluff: Students soon learn when you are trying to fool them. Acknowledge a mistake. The simple admission, "I do not know, but will find out", or "You were right and I was wrong" can do much to develop and maintain a good working relationship.

c. Be loyal: To your class, your superiors, and to existing policies. Correct at once any errors of administration affecting your students, but carry out the full extent of the teaching directive from your superiors.

d. Act decisively: In making a decision, give full consideration to all aspects of the problem and then act with conviction. State the approved position firmly and positively.

e. Abide by decisions: A fair decision, once made, should be carried through, and a propet order, once given, should be executed. Students will respect you for your decisiveness

f. Keep the student headed toward his objective: See that all activities of each lesson are directed toward achieving the established goal.

g. Respect the rights of your students: Always place your class ahead of your personal convenience and consider their rights as students. Be sensitive to how you emotionally affect others. Emotion is present in any job or teaching situation and may work for you or against you.

h. Be courteous: Correct student mistakes as they are made, but do so in a straightforward, impersonal manner. Never be sarcastic or make your criticisms personal.

i. Be cheerful and enthusiastic: The enthusiasm you have for your teaching job is directly reflected in an increase in student interest in learning and in a decrease in disciplinary problems. If you are cheerful, students will reflect your attitude. Know when to use humor in the classroom; good clean humor will create good will and help you fulfill your instructional mission.

j. Be business-like. Your job is important and there is no time for foolishness. Encourage initiative and self-reliance. A student who has learned to think for himself will be able to handle problems for which he has been taught no prescribed procedure. Welcome the assistance of students who have exceptional backgrounds, and use them as an aid in arousing the interest of less experienced members of the class. Each instructional problem is a cooperative learning experience.

4. PHYSICAL COMFORT.

   a. No matter how interesting your presentation, or how good an instructor you are, students will have difficulty paying attention if they are uncomfortable. Physical conditions constantly influence every minute of instruction. You must foresee possibly unfavorable conditions and take every action to make students comfortable whenever and wherever possible.

   b. Here is a physical environment check list:

   (1) Room temperature satisfactory?

   (2) Room well ventilated?

   (3) Chart or chalkboard glare eliminated or reduced?

   (4) Lighting where it is needed?

   (5) Distracting materials removed from walls?

   (6) Distracting outside noise eliminated or reduced?

   5. STUDENT SAFETY. The importance of safety practices cannot be overemphasized. Whenever you demonstrate a new skill to be learned, show students the reasons for safety practices. Safety consciousness is an attitude which must be developed in students. Insist that your students observe common sense precautions. Be sure that you have established all controls in your power over the human element that causes accidents.

G. CLASSROOM ADMINISTRATION

   1. GOOD CLASSROOM MANAGEMENT.

   a. The principles and procedures you adopt in handling necessary administrative details will contribute to your success or failure as an instructor. Start well before class time, check the learning area, classroom, shop, and problem area to be sure that all necessary equipment, supplies, and training aids are in proper condition and ready to be used.

   b. Some suggestions that will assist you in handling administrative details:

   (1) Begin and end class promptly.

   (2) Check attendance.

   (3) Make students responsible for routine details such as clean up of an area.

   (4) Check appearance and conduct of AI's.

   (5) Report all inadequacies of equipment: i.e., sound, classrooms, training aids, or heating.

   (6) Do not permit students to "close shop" during the final summary.

Chapter VII

TRAINING AIDS

A. TYPES AND SPECIAL OPERATION TECHNIQUES

1. GENERAL.

  a. Training aids, as the name implies, should be used to aid understanding--to assist the student in learning. When you present an idea to students by means of words alone, students must picture in their minds what you are trying to convey through these words. Psychological research has demonstrated that most students learn more easily through the sense of sight than through any other sense. Realizing this, capitalize on this principle by using visual aids whenever they will help to present the subject in a more understandable manner. Training aids will make instruction more meaningful by reinforcing the student's sense of hearing with an appeal to another sense. Training aids will not take the place of verbal explanation but they will help to make such explanation clearer to your students.

  b. The object of this chapter is to explain the purpose, types, and value of training aids. The techniques of proper use and future trends in training aids will also be discussed.

2. PURPOSES.

  a. The major purposes of training aids are: to reinforce explanation, to direct student thinking to a specific item, to aid retention, and to save training time.

  b. Training aids reinforce explanation by appealing to many of a student's senses simultaneously. Besides hearing your explanation, students can see what you mean. Through use of both senses clearer understanding will result. For effective instruction all students must have a comprehensive mental picture of the subject being discussed. One way you can make sure that all students receive a similar mental picture is by using a training aid which helps to standardize the visual picture.

  c. Training aids focus student thinking to a specific item. You know that when a visual training aid is being discussed, explained, or demonstrated, students are concentrating on the aid because objects which appeal to the sense of sight have great power in holding student attention. When you are lecturing, students may appear to be listening but they are possibly thinking of many other things.

  d. A good training aid assists student learning by increasing retention. Since impressions secured through visual means are more lasting and vivid than those acquired through hearing, visual aids play a dominant role in the learning process by insuring greater retention. The greater the impact of learning on our senses, the deeper the impression with increased chances of retention.

  e. Training aids frequently save time and replace long involved explanations by the instructor. It is virtually impossible to explain some technical ideas or processes by means of words alone. A good chart, an operating model or mock-up will put across the ideas more effectively and more quickly. When training operates within a compressed time schedule, any device which speeds up average learning time is extremely valuable.

3. SELECTION OF AIDS.

  a. Let's assume at this point that you have done the research and organizing necessary to complete your lesson outline. Your outline is apparently all set, but there are several teaching points or instructional areas which are complex and difficult to explain by words alone. You remember that the function of a training aid is to assist in increasing student learning and retention of material.

    b. The first step in selecting training aids is to sit down and carefully examine the instructional objective and teaching points because a training aid must definitely aid the students in understanding this objective. If it fails in this respect, it ceases to become an aid and becomes a distraction.

    c. There is a limit to the amount of material that a student can understand at any one time even with the help of accurate training aids. Too many ideas presented at one time or too many details on any one training aid will create confusion and hinder student learning. It is far better to present a few simple ideas on a chart and to use several charts than to attempt to crowd everything into one aid. In planning training aids simplicity is an important word to remember. A chart is not supposed to be a condensed version of your teaching notes. New instructors frequently err in trying to place too many notes on a chart or projectual. The aid should contain only the key noun, verb, or phrase. It is not designed to stand alone without explanation. However, words or phrases will never have the impact upon student senses, the gateways to the mind, that an appropriate picture or visualization would. Strive to visualize the concept in your training aids. Too many charts and projectuals are merely lists of words that could just as easily be written on the chalkboard by the instructor during the class; this would be more interesting and dramatic. Don't make a slide to use as a memory crutch.

    d. If your instructional objective is to familiarize your students with the operation of the gas system, one of the best aids would be a large schematic drawing of the system, but you would not want to use this as the sole means of getting across your objective. You could project slides to show parts of the schematic or even bring in a training film that explains how it works. Perhaps a schematic model would help to explain some difficult points. Keep in mind that you can use various types of aids to explain the same idea. By using a variety of aids during the period instead of all charts, all slides, or all film, you increase student interest. Of course, there is a point of diminishing returns here as well, since you do not want to make your instructional hour like a circus - flipping from one type of aid to another indiscriminately.

    e. Careful evaluation of your instructional problem will reveal where a chart, model or other training aid will promote better student understanding. Remember that there are many types of aids and each has certain advantages and limitations. Frequently it will be necessary to use more than one type of aid to present a lesson effectively. It is of utmost importance to weave the aid into your lesson pattern carefully. The actual use of the aid must not interrupt the flow of ideas during the lesson presentation. To be effective, training aids must be a natural, intrinsic portion of the sequence of instruction.

    f. A good rule of thumb to follow when determining whether to use a training aid or not is to ask yourself, "Could I achieve the lesson objective as well without the aid?" If the answer is yes, then either you have selected a poor aid for your subject or no aid is necessary at that particular time. Remember, training aids are not "crutches" for you to lean on; they are incorporated in the lesson to assist student learning. They supplement instruction, they do not replace it. The criterion, "Does it help student understanding?", is the main standard by which the training aid should be measured. The other advantages such as "to appeal to more than one sense", "to focus student attention", "to increase student interest", apply to a good training aid but should never be the main consideration.

4. <u>TYPES OF TRAINING AIDS</u>.

    a. <u>Variety of Training Aids</u>. Training aids in use include actual equipment, models, charts, chalkboard illustrations, training films, other types of projectuals, and recordings.

b. Actual Equipment.

(1) The actual equipment is the most realistic training device. It always creates a lasting impression to have the actual object present during instruction. However, sometimes the actual object might be unavailable or inappropriate for class use. It may be too large or too small for formal classroom instruction.

(2) At times it may be desirable to have both the actual object present in combination with other training aids to illustrate the object. Certain features can be shown on a model or chart that cannot be shown with the actual equipment since they may be internal portions or features of the object. Study your lesson to see if it is possible to use the actual object in your period of instruction. Remember that if the weapon is before them, students will show greater interest and they will certainly gain a better understanding of what you are teaching.

c. Models (training devices).

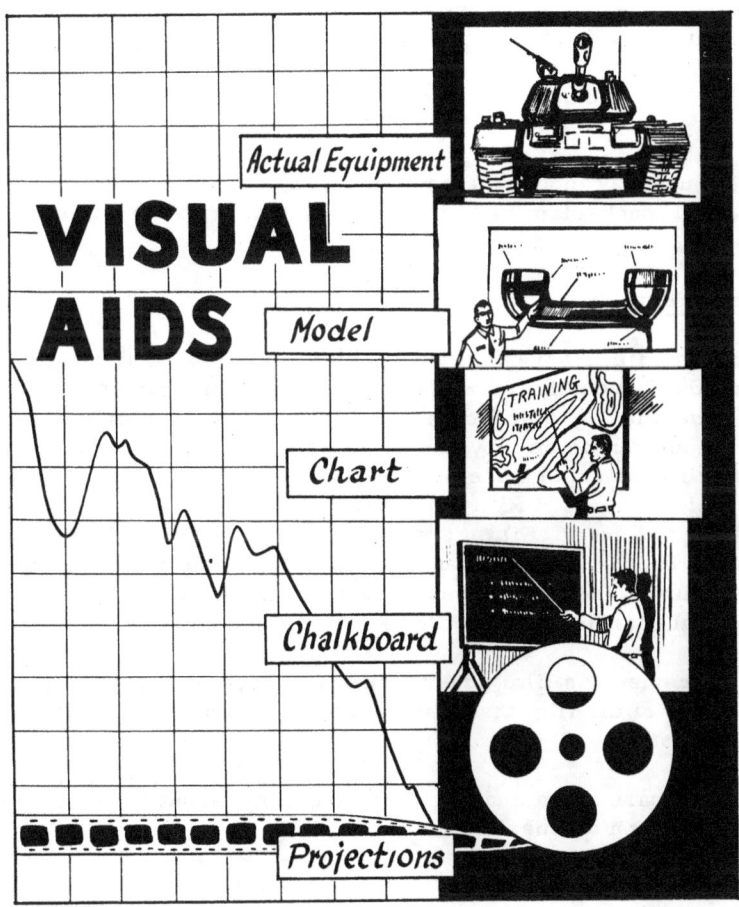

Figure 7-1. Visual Techniques.

(1) Models are of little practical value unless they are made to scale and give an accurate representation of the actual object. They must be of suitable size for the teaching purpose and they must be convenient to handle. Above all, they must be appropriate and necessary to the fulfillment of the teaching objective. A model suitable for a ten-man class may not be effective for a 50-man class.

(2) There are many different sizes and types of models. Analyze your training situation carefully before you decide the type of model to suit your purpose. Here are some of the more common types:

(a) Actual size model - a model of actual equipment made to exact scale. One reason to use actual size models in lieu of actual equipment is that the model may be made of lighter material or not necessarily operable; hence it is much cheaper for use in instruction. You don't have to tie up the actual equipment when the model would serve the purpose. It can be field stripped exposing parts not otherwise accessible.

(b) Enlarged scale model - one made to a scale larger than the actual equipment. For example, an enlarged model of the sights to instruct classes in the correct methods of obtaining a correct sight alignment.

(c) Reduced scale model - here the model is made to scale but smaller than actual equipment. Usually this type is used for small group work and is more portable and storable than the actual equipment. Frequently, for range procedure classes, a reduced scale model of a range may be used.

(d) Working model - a model where the parts move to simulate the actual operation.

(e) Cut-away model - a sectionalized or cut-away view of a piece of equipment to display the internal mechanism in such a manner that students can see the parts and operation.

d. Training Films.

(1) Training films have won a well earned place in instruction; when properly designed and used, they are extremely interesting and vivid training aids. By using the training film an instructor can bring into the classroom all of the impressions of outdoors. With training films, specific situations can be employed which would otherwise be difficult or impossible to present. Films can bring the experience of experts into the classroom to help explain difficult portions of your problem. A good film will assist student learning and save training time when carefully planned for and used. Films are excellent for both standardizing instruction and accelerating learning, but there never was a training film that could replace a well-prepared instructor. Don't use a film as the sole means of instructing but as an aid to your teaching. Before you use a training film, carefully follow these steps:

(a) Preview the film several times to become thoroughly familiar with its content. Select the portions of the film that you wish to show and do not show irrelevant portions, no matter how interesting.

(b) Prepare notes and pertinent questions about the film to use in class orientation and post presentation discussion.

(c) Introduce the film to students carefully, directing attention to the purpose of the film and the specific items to look for.

(d) Show the portions of the film as planned. Remain in the classroom while the film is being shown.

(e) After showing the film, review the main points which it brought out. This summary will reinforce the teaching ideas and increase the instructional value of the film. Answer student questions and ask questions to check student learning.

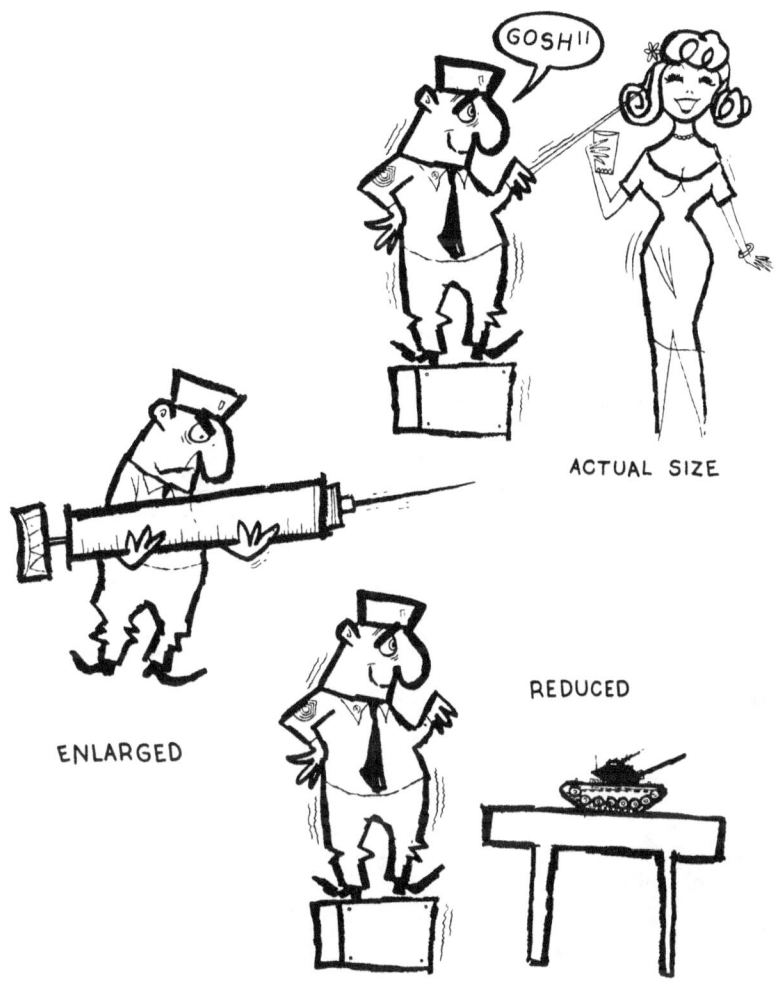

Figure 7-2. Models Are Excellent Training Devices.

(f) If time permits, reshow portions of the film. This is an extremely valuable step which is all too often overlooked by instructors. Experimental studies have demonstrated that students get much more out of the second showing of the film than the first showing. It is amazing how many important details students overlook during the first showing.

(2) Training films have certain disadvantages which you, as an instructor, should be aware of and guard against. When they are projected front view, you must darken the room. This creates an uncomfortable, hot situation in non-airconditioned classrooms. Since there was a time when training films were overused in mobilization training, many students were unfavorably conditioned against films and they look upon film time as a good time to sleep. Never show a film simply to entertain students - never show portions of a film that are irrelevant to the teaching points being considered. When used sparingly, they are an excellent training aid; when overused they are a drug.

e. Film Strips.

(1) A film strip consists of a length of standard motion picture film containing still pictures of a specific subject. A film strip is most effective when a series of related pictures must be presented in a definite order or sequence. A good example of a film strip use would be the disassembly or assembly of a mechanical unit such as a M14 or 45 cal pistol. An

advantage in using a film strip is that you operate the strip projector yourself which permits a personalized approach. Again, you control the rate of presentation and can project the various pictures to dove-tail with your words. If there are any student questions, the frame can be re-shown and discussed. The sequence can be stopped at any time to discuss and to clear up points not understood.

(2) The main disadvantage of using film strips is that it is generally necessary to have the room blacked out and the blinds closed. This reduces personal contact between instructor and the class and may encourage sleeping.

f. <u>35MM Slide</u>. The 35mm slide or film strip projector is becoming more and more popular as an instructional aid. 35mm slides are not so expensive to make or reproduce as large wall charts. Yet they serve the same purpose. Using a 1000 watt bulb you can show a number of pictures, graphs, maps or diagrams projected up to 18' x 26' in size at 100 feet distance and yet you carry the slides to class in your pocket. A disadvantage of front view projection of the 35mm slide projector is that the room has to be darkened. This hampers students from making notes readily and you have difficulty in seeing group and individual reactions so you miss that on-the-spot evaluation of how your instruction is progressing.

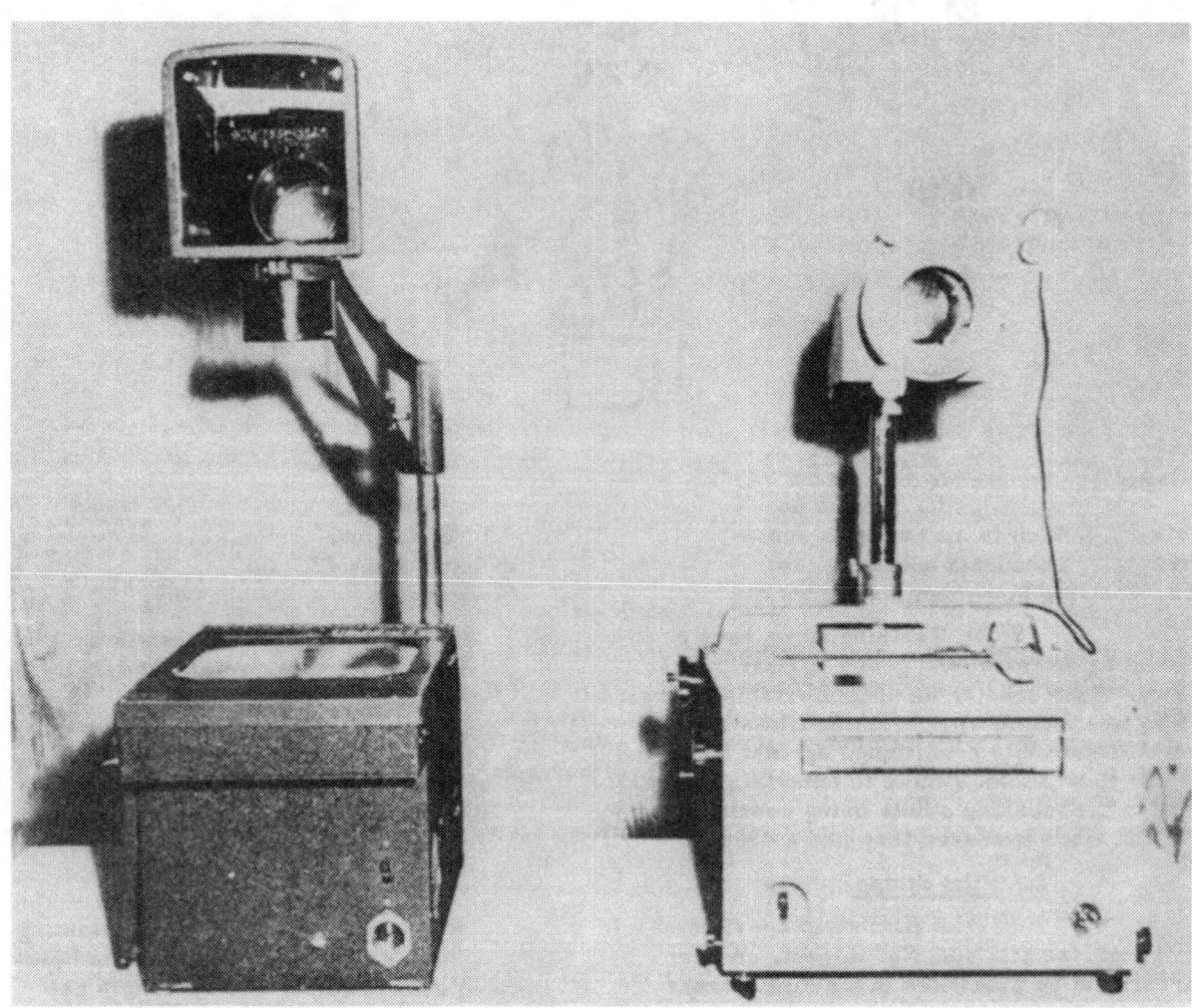

Figure 7-3. Overhead Projectors.

g. Overhead Projector.

(1) The overhead projector accommodates clear 10" x 10" acetate slides. The overhead projector is a fast, convenient and flexible means of projecting instructional material. As such, it has earned its rightful place as one of the most useful training devices at USAIS. With this projector you can use as many charts or map projections as you wish. You can display material quickly and easily and expense and effort necessary to prepare slides is not great. Color can be used on transparencies to add interest. A single mounted transparency can be reproduced in several layers with each layer containing several colors.

(2) Another variation of the overhead projected transparency is the "flip over" transparency. Here a number of transparencies are hinged to the side of a single mount and each side is flipped on as you call it. With the 1000 watt projector you have as many as six or eight flip overs without materially affecting the quality of the picture. A special situation can be built up very effectively by means of such flip overs. It assists sequential or step-by-step presentation of learning materials.

(3) The major advantage of the overhead projector over other types of projector is that is can be used in lighted classrooms or normal daylight conditions. Students can make notes and you can observe their reactions. The projector is excellent for formal or informal instruction. Its few movable parts make it extremely serviceable and easy to maintain.

h. Wall Charts.

(1) The wall chart is a drawing or sketch on medium heavy drawing paper which generally represents a schematic or topical outline of ideas. A well designed and skillfully used chart may frequently make the difference between a highly effective and interesting lesson and one that is dull, uninspired or confusing. Sometimes a wall chart is preferable to a projected aid especially if extended discussion is to follow each point, because the use of a wall chart eliminates the problem of a partially darkened room with attendant lack of ventilation and student attention.

(2) Wall charts help to build ideas or concepts cumulatively and assist the logical thinking of students. They add visual reinforcement since they appeal to another student sense. One good technique is to develop the fundamental ideas by means of the chart, then put the chart away as the conference or discussion develops and lastly, to reuse it in the summary to reemphasize the fundamental ideas of the lesson.

(3) Each classroom should be equipped with chartboard panels. In placing your charts on these panels be sure that the lectern or curtains are not blocking the view of the charts. Pay particular attention to the centering of the chart on the panel. If the chart paper has been cut squarely, you correctly center it by keeping the distance from the edge of the chart to the edge of the sliding panel uniform. Every chart should have a simple, yet comprehensive title describing the subject material portrayed on the chart. Simplicity is the keynote for good charts. Work deliberately to reduce your teaching points to brief, simple language; don't clutter the chart with unnecessary words or details. Put only the essential facts on the chart. Phrases on a chart are usually condensed to keywords. The chart is not expected to teach the whole lesson for you and is not expected to be completely understandable without explanation by the instructor. It is an aid, not a copy of the instructor's notes.

(4) There are different ways to focus student attention to an item on your charts. One way is to use a pointer to point out specific ideas being discussed. Another interesting way is the "strip tease" method. With the "strip tease" method you conceal portions of the material by cover strips until you actually need them. When you are ready to proceed to the new point in your lesson, you remove the next strip from your chart to reveal the next point. Still another variety is the venetian blind technique where you open each slat as you develop that idea.

(5) Key words sometime assist student learning and make charts more interesting. In this key word technique you use the first letter of the first word of each idea. Sometimes this first letter is printed in a different color. When read vertically the letters should make some simple English word. If the word ties in with your subject matter so much the better. It will reinforce student memory. Use a key word only when it will definitely assist learning. Use key words only where rote memory of a sequence is desired since many times key words tend to confuse students and many instructors tend to overuse this memory device.

i. <u>Chalkboards</u>.

(1) The most valuable training aid at this or any school is the chalkboard. All classrooms should be equipped with at least two chalkboard panels and portable boards available for outdoor instruction. Hence a chalkboard is always available for instructor use. It is a flexible, inexpensive method of presenting ideas and students will learn more readily when they are able to see as well as hear.

(2) Another advantage of the chalkboard is that it is an excellent way to promote class participation by helping the instructor and the class develop the subject together. It is an ideal aid for the conference method of instruction. As the class gives various comments, suggestions, or answers to your questions, you place their ideas upon the chalkboard and students feel that they have contributed to the class. Then as you summarize at the end of the period you can use the ideas they gave you which you wrote on the board.

(3) A chalkboard enables you to make any picture or sketch you wish or you can hand-make your own training chart on the spot by preparing the board in advance. A useful technique that many instructors use to reproduce material on the chalkboard is to project the material from an overhead projector directly on the board, then copy in pencil the important lines. These pencil lines will not be seen by your class but you can copy over them in chalk when the time comes for developing them.

(4) Here are some suggestions regarding the proper use of a chalkboard:

(a) Plan your chalkboard work carefully in advance. Take into consideration such things as: the amount of space available, proper layout of your material, and adequate visibility by the most distant students.

(b) Pencil in advance any required sketches on the chalkboard. Here you may use a straight edge and any other aid in a calm and precise manner so that when you copy over in chalk the sketches will be of fair quality despite the fact that you will be chalking in rather hastily during the actual class.

(c) When you put material on the board, print in large, legible block letters. Use only one type letter throughout and make your printing neat and legible.

(d) Put your material on the board in a simple brief manner but be sure that it is complete enough to be understood by the average or weak students. Abbreviations may be used as long as they are the usual standard abbreviations. Don't use too many abbreviated words one after the other or students may lose the sense of the idea.

(e) Some instructors use the wet board technique very effectively. Here you get the board very wet and then write in soft chalk. If you time your presentation accurately as the board begins to dry, the key word or ideas appear as if by magic on the board. Timing of this method is rather tricky but when properly used the method is effective. A variation of this is the "strip tease" method discussed under wall charts; paper strips can be used on chalkboards very readily.

(f) Don't erase until all students have had time to see the material and copy it if they desire.

(5) There is no more dramatic way of presenting material visually than by use of the chalkboard. Persons love to see things or ideas developed before their eyes since it stimulates their creative instincts. Remember how often as a child you watched a cartoonist at work in such movie programs as "Out of the Ink Well" or how you watched spellbound as sand or snow sculptors developed their material. Writing teaching points on the chalkboard as you secure them from student discussion is doubtless the most effective way of conducting an instructional conference for the student feels that everything has not been canned or worked out in advance but rather that this lesson was a result of a stimulating interchange of opinions between instructor-student and among various students.

j. Recordings.

(1) Any instructor may make tape recordings for his problems. These recordings may be played directly through the sound system installed in the classroom or through portable sound systems for outdoor problems. Such tape recordings will add drama, interest and variety to your instruction and increase student learning.

(2) By using recordings you can introduce other voices into the classroom to add interest and authority to the presentation. There is a dramatic appeal to the recorded word since your classes have been conditioned to radio and television listening. Recording the skit will save repeated use of AI's and other instructional personnel as you present the problem time after time.

(3) Recordings force the student listener to recreate the situation and to use his imagination. This frequently appeals to many students who have a sense of the dramatic. Be careful not to overuse recordings since what we learn through our ear is not so readily understood and retained as what we learn through seeing. When you use imagination in your teaching through bold and dramatic training aids, student learning is stronger since the initial impression is deeper.

B. PROPER USE OF TRAINING AIDS

1. GENERAL. After selecting the desired aids, you must know how to use them properly. Nothing destroys class morale more than watching an instructor fumble with a chart not properly displayed, struggle with equipment not ready for use, or discover at the last minute that all of his training aids are not at the class site. Follow these general rules in using any type of training aid:

a. Preparatory Preview. In this step make sure that all of your aids are clean and in good repair. Remember, a shoddy training aid indicates a careless instructor. When you have training aids made, check the content of charts and transparencies carefully for possible mistakes in spelling and grammar. Because you submitted a perfect copy does not mean you don't have to recheck. The time to correct errors is before the class begins and not while you are presenting the problem. Explain carefully to your AI what you want him to do while you are using this training aid; where you want it displayed; when you intend to use it; and how you want him to help. Rehearse this with him to insure smoothness.

b. Introduce the Aid Skillfully. Since a training aid is used to assist learning, some discussion should precede the showing of the aid to give a logical reason why a student should look at or listen to this aid. Too many instructors simply flash an aid before a startled class and begin talking about it without properly introducing the subject and conditioning the students to the aid. The aid should appear at the strategic time because student need for the training aid has been built up by proper introduction to it.

c. **Show the Aid.** Display your training aid smartly and skillfully without undue distraction. Rehearse the use of aids before your presentation. Show the aid so that the entire class can see it. Position yourself to avoid obstructing student view. Use a pointer, a spotter or the strip method to focus attention on the specific portion of the aid under discussion. While showing the training aid talk to your class and not to the aid. Generally, it is not good to talk when revealing or putting away an aid.

d. **Review the Aid.** Frequently the question is asked, "How many times should I show a particular training aid during a one or two hour problem?" Show it as many times as it is necessary to assure yourself that the class understands and that you have reached your instructional objective. Many instructors give a quick summary by reshowing their training aids near the end of each problem. Some problems may require that the training aid continue to be exposed during the instructional conference but it is generally a good practice to put away the aid or cover it as soon as you are done with it.

CHAPTER VIII

REVIEW OF COURSE OF INSTRUCTION AND WEAK POINTS

A. THE REVIEW OF THE COURSE will be covered as outlined in Lesson Plan (I- 8) "Review of Course of Instruction." Page 204

B. WEAK POINTS in the shooter's performance are subject to the extra supervision of shot analysis and positive corrective measures. Various performance flaws peculiar to slow fire techniques are discussed in Chapter VI, Slow Fire Techniques, Advanced Pistol Marksmanship Manual. Other shooting deficiencies peculiar to rapid fire technique discussed in Chapter VII. The following weak points are deemed the most frequently observed by USAMTU coaches in the course of thousands of man-hours of advanced pistol marksmanship instruction.

1. IMPROPER GRIP: Natural sight alignment depends upon immobility of the head and a proper uniform grip being maintained throughout all firing. Proper grip, as determined by the initial check-out, cannot be maintained as such, without frequent rechecks to insure its continuity. Resist the tendency for incorrect grip to become a built-in error that will undermine your performance.

2. IMPROPER POSITION: An improper position will cause the weapon to settle to the right or left of the bull's-eye instead of directly underneath or directly on it. Vital seconds in timed and rapid fire are lost by trying to correct the error in hold caused by faulty position after each shot is fired. The proper position will tend to cause the pistol to point naturally at the center of the aiming area. The built-in error associated with faulty position is of equal consequence to that of error in grip.

3. EARLY FIRST SHOT IN TIMED AND RAPID FIRE: In rapid fire it is necessary to break the first shot immediately after the target turns, to eliminate worry about time and to maintain a definite rhythm. An early or late shot may destroy all chance of posting a winning score.

4. LACK OF RHYTHM IN TIMED AND RAPID FIRE: A definite rhythm or cadence is a must, especially in rapid fire, where only 10 seconds is allowed. After recovery, continue a positive, uninterrupted, pressure against the trigger, keeping the sights as perfectly aligned as possible. Hesitation in rhythm, while trying to correct any detectable error in hold during rapid fire, breaks rhythm, concentration and loses critical time.

5. LACK OF FOLLOW THRU AND NATURAL QUICK RECOVERY: Follow thru is the continuation of all mental and physical processes brought into play to deliver a surprise break and a controlled shot. The effort is sustained beyond the point where recoil indicated that the bullet is on the way to the target. The prime indication that it is a surprise shot is the lack of reflex action of the arm or hand muscles that disturbs sight alignment. Recovery is the return and settling of the weapon into the center of the aiming area, after recoil has taken place. In rapid fire, recovery should be accomplished as quickly as possible with center hold and perfect sight alignment naturally reestablished.

6. HOLDING TOO LONG IN SLOW FIRE: A limited amount of oxygen can be momentarily stored in the respiratory system and the breath can be held comfortably for only a short length of time. It stands to reason, that within the first ten or twelve seconds the shooter is in better condition to deliver a good shot on the target. Waiting too long to apply a positive trigger pressure to break the shot may allow the arc of movement to increase and/or the eye focus to blur. The shooter eventually becomes impatient and abruptly speeds up his trigger action. This action causes a jerked shot.

7. LOOKING AT THE TARGET: The eye can only focus on one object at a time. That object should be the front sight. Valuable time is lost in looking over the sight to the target to adjust the point of hold. During the fleeting instant that you remove your focus from the sights, they could become misaligned. If trigger pressure continues and the shot breaks at this time, the bullet may not strike in the desired area. Most shooters have at one time or another experienced this occurrence. It is difficult to maintain constant point focus on the front sight, but it can be effectively accomplished through continued emphasis on training in this specific area.

8. ANTICIPATION: The reflex action of heeling one o'clock hits and the changing of the the characteristics of your arc of movement, from normal to a straight line movement, are the symptoms of anticipation. The human body has a natural tendency to brace itself against an expected shot. This is caused by a subconscious awareness by the shooter that he has decided to moderately speed up the trigger pressure, because conditions are ideal NOW and may not remain so for very long. Muscular and/or nervous reflex action takes place and disturbs sight alignment because the reflexes usually react sooner than the actual firing of the shot. The instant nature of the reflex action sometimes so closely coincides with recoil that difficulty is often encountered in making an accurate shot call.

9. IMPATIENCE-JERKING: A spasmodic, abrupt trigger movement combined with the reflex action of the arm and hand muscles causes a violent disturbance of a sight alignment and results in a bad shot. When this happens, the shooter may be attempting to use point shooting technique but faulty trigger control prevents the application of smooth trigger pressure. Examples of possible conditions:

    a. When the shooter has perfect sight alignment and is able to momentarily freeze the arc of movement. The condition is of extremely short duration and the shooter is prone to act quickly and abruptly in applying trigger pressure before the ideal moment passes.

    b. Attempting to fire the shot during an enlarged arc of movement, such as in wind shooting and initiating a squeeze at the instant the aligned sights move across or move into the vicinity of the aiming area.

    c. After holding too long, many shooters become impatient and exert abrupt pressure on the trigger. The knowledge that he will be unable to sustain his present satisfactory minimum arc of movement in the center of the aiming area prompts the spasmodic trigger action.

10. COMPLETE SHOT ANALYSIS is necessary for improvement of a shooter's performance and when errors are made, corrective action should be taken immediately. (Example: If the shooter is heeling or jerking, he should stop and dry fire several rounds to regain smooth coordination of employing the fundamentals.) Use the shot analysis guides on the slow fire or rapid fire worksheets.

11. DO NOT SHOOT AND PRACTICE ALONE: When possible, shoot with someone who is a better shot than you. Observe his methods, the sequence of events he uses and request that he coach you for a few strings. Most good shooters are willing to help another shooter to perfect his performance.

12. UNRESOLVED SHOOTING PROBLEMS handicapping performance are a continuing burden plaguing a shooter on the firing line. As a competent shooter knows, persistent shooting faults mean a slump in performance. As a result, he is not in the proper mental state when on the firing line and cannot concentrate properly. His anxiety is that these faults will continue to affect his performance and will handicap his scores. A so-called slump need only last until planning for the next shot. The solution of those knotty problems in your shooting can be best resolved by comprehensive shot analysis and application of corrective techniques. A good line coach or one of your teammates can be of great help to you in your dilemma.

13. <u>DON'T KNOCK IT UNTIL YOU HAVE TRIED IT</u>: Suggested techniques should be fully explored and tested. The old saying that "Rome wasn't built in a day", applies here. When an expert suggests changes in any part of a developing shooter's performance, the shooter should give the new method a thorough trial before discarding it. The master shooter attained his rating by doing many things right.

<u>CONCLUSION</u>: The knowledge you have gained in analysis and corrective technique will remove the limitations on your performance. No record score is sacred. No champion is inviolate. The deliberate and cool efficiency with which you deliver countless controlled shots to the ten ring will pay off in combat situations where this same efficiency will kill with ruthless regularity.

Chapter IX

EXAMINATION AND CRITIQUE WITH PANEL DISCUSSION

A. WRITTEN EXAMINATION. The use of tests or examinations to evaluate student performance is a necessary step in the teaching process. Instructors must use tests to determine overall training progress and they must also use tests to check on the effectiveness of instruction. It cannot be assumed that men have learned until the examination stage of instruction has revealed a desirable standard of achievement.

1. PURPOSES OF TESTS OR EXAMINATION.

    a. Tests aid in improving instruction by-

        (1) Discovering gaps in learning. Properly constructed tests reveal gaps and misunderstandings in student learning. If frequent tests are given, such weaknesses can be discovered and instructors can correct them by reteaching their material.

        (2) Emphasizing main points. A test is actually a valuable teaching device in that students tend to remember longer and more vividly those points which are covered in an examination. Tests encourage students, as well as instructors, to review the materials that have been presented and to organize various phases of instruction into a meaningful set of skills, techniques, and knowledge.

        (3) Evaluating instructional methods. Tests measure not only student performance but also instructor performance. By studying the results of tests, instructors can determine the relative effectiveness of their various methods and techniques.

    b. Tests provide an incentive for learning. Students learn more rapidly when made to feel responsible for learning. For example, they are more likely to pay close attention to a training film if they know a test will be given when the picture is over. Generally, instructors who more frequently give tests will find that their students will be more alert and learn more. There is a danger, however, in overemphasizing tests and test scores as a basic motivation for learning. Student interest in test scores is a superficial one which can easily lead to efforts to "hit the test" rather than learn the subject matter for its value in the future. Students who study primarily to pass tests may forget what they learn much faster than those who are interested in learning because of the real value to be derived. The instructor should give rigid tests and give them frequently, but they should be designed to require the student to make application of what he has been taught.

    c. Tests provide a basis for evaluation. Another purpose of testing is to determine which students have attained the minimum standard of performance and which have not. In many cases it is desirable to indicate the extent to which students exceed or fall below the standards required. Students learn different amounts; the grade recorded for each student should be an accurate index of what he has learned. Unless a sound testing program is employed, it is impossible to determine the relative achievement of students.

    d. Tests furnish a basis for selection for further advanced training. The results of training tests furnish valuable supplementary information if tests are actually a measure student performance. The test results become a valuable basis for determining whether a student should be placed in another activity or whether he should receive advanced training.

2. CHARACTERISTICS OF AN EXAMINATION. There are six important factors which affect the quality of an examination. These factors, while not considered to be separate and distinct, are defined and discussed separately in order to develop a clear understanding of the characteristics of an examination.

   a. <u>The test must be valid.</u>

   (1) The test must measure what it is supposed to measure; this is its most important characteristics. A test designed to measure what students have learned in a specific training program should measure achievement in that training program and nothing else.

   (2) The instructor should, whenever possible, invite the opinion of other competent persons as to the validity of his tests. The test results obtained should be compared with other measures of student achievement. A variety of tests and other evaluating devices must be used in obtaining a valid measure of achievement.

   b. <u>The test must be reliable.</u>

   (1) A test is said to be reliable when it measures accurately and consistently. If the test measures in exactly the same manner each time it is administered, and if the factors that affect the test scores affect them to the same extent every time the test is given, the test is said to be highly reliable. This characteristic of a test is especially important when tests are used to compare the proficiency of several classes.

   (2) There are several factors which affect the reliability of a test. In general, the reliability of a test can be raised by increasing its length. The more responses required of students, the more reliable is the measurement of their achievement. Test items should be designed to make it difficult to guess the correct answer. Also, the way in which a test is administered, and the conditions under which it is given should be consistent.

   (3) Other characteristics of the test, such as validity and objectivity, also contribute to its reliability.

   c. <u>The test must be objective.</u> A test is objective when instructor opinion, bias, or individual judgment is not a major factor in scoring it. Objectivity is a relative term. Some tests, such as written examinations which are machine graded, are highly objective; others, such as essay examinations, written exercises, and observation techniques, are less objective. Sometimes observation is the only effective way of determining proficiency. In such cases the instructor must strive to make his observations as objective as possible.

   d. <u>The test should discriminate.</u> The test should be constructed in such a manner that it will detect or measure small differences in achievement or attainment. This is essential if the test is to be used for ranking students on the basis of individual achievement or for assigning marks. It is not an important consideration if the test is used to measure the level of the entire class or as an instructional quiz where the primary purpose is instruction rather than measurement. As is true with validity, reliability, and objectivity, the discriminating power of a test is increased by concentrating on and improving each individual test item. After the test has been administered, an item analysis can be made which will show the relative difficulty of each item and, of greater importance, the extent to which each discriminates between good and poor students. Often, as with reliability, it is necessary to increase the length of the test to get clear-cut discrimination. Three things will be true of a test that has discrimination:

(1) There will be a wide range of scores when the test is administered to the students who have actually achieved amounts that are significantly different.

(2) The test will include items at all levels of difficulty. Some items will be relatively difficult and will be answered correctly only by the best students; others will be relatively easy and will be answered correctly by most students.

(3) Each item contained in the test will possess discrimination. If <u>all</u> students answer an item correctly, it is probably lacking in this respect.

e. <u>The test must be comprehensive.</u> It must sample liberally all phases of instruction which are covered by the test. It is neither necessary nor practical to test every point that is taught in a course; but a sufficient number of points should be included to provide a valid measure of student achievement in the complete course.

f. <u>The test must be readily administered and scored.</u> It must be so devised that a minimum amount of student time will be consumed in answering each item. The test items must also be constructed so that they can be scored quickly and efficiently.

3 TYPES OF TESTS.

    a. Written.

    b. Oral.

    c. Practical exercise or performance.

B. CRITIQUE

1. A STAGE OF INSTRUCTION.

    a. <u>The critique is the final stage of instruction.</u> It is designed to review the lesson and reemphasize the teaching points. An examination or practical exercise should always be followed by a critique, otherwise students may not have a clear, orderly idea of what was done right and what was done wrong. Good instruction includes intelligent, tactful, and constructive criticism; this criticism can be given most effectively in a group discussion held after an exercise or examination. The critique can be used to -

(1) Sum up and clarify a situation developed in the lesson or test and point out correct or incorrect methods of execution.

(2) Provide an overall view of the entire applicatory operation.

(3) Indicate the strong and the weak points of a performance and methods or procedures to be used in correcting errors or mistakes.

(4) Reemphasize the fundamental points of the lesson.

(5) Develop among personnel a spirit of unity and an appreciation of the cooperation and teamwork necessary in team activities.

    b. <u>The critique is so important</u> that it must be considered a stage of instruction in itself. However, it is most valuable when it becomes, in effect, a part of another stage. Every period of presentation, application, and examination should include a well-integrated critique. The effectiveness of this stage depends upon the flexibility with which the instructor employs it.

c. Human Relations Are Important. In conducting a critique, the instructor must not be sarcastic; he must make criticisms or comments in a straightforward, impersonal manner. If deemed necessary, he should criticize individuals in private, praise them in public. Students should leave the critique with a favorable attitude toward the training and have a desire to improve.

d. The Critique Should Relate the Instruction to the Subject or Course. It should emphasize the continuous nature of training by calling attention to what has been done earlier and to the relationship of the instruction just completed to the subject or course of which it is a part.

e. Specific Points Should be Covered. Procedures used, examples of personal initiative, type of errors and ways for correcting them, and fundamental teaching points should be covered specifically.

f. Fundamentals Should Be Emphasized. The critique should indicate the various acceptable solutions; it must not give the impression that there is but one correct method of solving the problem. Such a misconception leads to the adoption of stereotyped solutions and to attempts to guess the approved solution, resulting in loss of initiative and independent thought. The critique should emphasize the fundamental principles and should criticize and evaluate the different solutions on the basis of their completeness, effectiveness, and observance of the fundamentals.

g. Participation Should be Encouraged. In almost every class there will be individuals who can contribute experience that will emphasize and illustrate key points. Too, a well-controlled class discussion makes the students feel that the critique is a period for learning rather than a time set aside for criticism of their performance.

h. Instruction Should be Foremost. The critique must be conducted as a stage of instruction and part of the course. A good critique might be said to "nail the lid" on the store of knowledge the student has gained during the course of instruction.

i. Steps in the Conduct of the Critique. The critique cannot be planned as thoroughly as other stages of instruction because the points to be covered are influenced directly by the members of the class. Advance planning can include the general outline; during other stages of instruction the instructor can take notes to guide his critique; but detailed planning is not practical. However, the instructor can insure complete coverage of the essential elements by following this general procedure:

Step 1: Restate the Objectives. This will enable the class to start its consideration of the instruction on a common ground. This step is necessary because some students may have become concerned only with a particular aspect of the subject and may have forgotten the overall objective.

Step 2: Review the Teaching Points. In this step, briefly summarize the methods used, or the teaching points brought out, and answer the student questions: "What was this all about?" "What did we do?" "How was it done?"

Step 3: Evaluate Performance. This is the most important part of the critique. Using notes taken during the exercise, the instructor points out and discusses the strong points. Then he brings out the weaker points and makes specific suggestions for improvement. He must be careful not to "talk down" to the group and must not expect a standard of performance beyond the capabilities of the students, considering their state of training. All remarks must be specific; students will not profit from generalities.

Step 4: Control the Group in Discussion. The instructor should encourage the class to discuss the points he has mentioned and to suggest other points for discussion. All the techniques of conducting a directed discussion apply in this step to insure that criticism is constructive and that discussion is to the point.

Step 5: Summarize. The critique should be concluded with a brief but comprehensive summation of the points brought out. The instructor can reemphasize teaching points and suggest specific practices to overcome certain deficiencies. The critique should be business like; it must not degenerate into a harrangue.

C. PANEL DISCUSSION. A panel discussion may be used in conjunction with the critique to provide greater latitude in subject coverage. The panel will be composed of the instructor-coaches and shooters who presented the course of instruction and various members of prominent pistol teams who by their experience and knowledge of advanced pistol marksmanship can contribute materially to the progressive understanding of all phases of the technique of control of the employment of the fundamentals of advanced pistol marksmanship.

ANNEX TO SECTION ONE, "TECHNIQUE OF PISTOL MARKSMANSHIP INSTRUCTION"

INSTRUCTIONAL GLOSSARY

ADVANCE ASSIGNMENT: Study or other work required to be performed by students prior to class. The advance assignment may be prescribed in the schedule or included in an advance sheet issued prior to the presentation.

ARTICULATION: Production and combination of separate sounds to produce intelligible speech. You violate good articulation when you run words together or fail to sound all parts of the word.

BODY (OF LESSON): The main part of a lesson, the lesson body, contains each student performance objective with appropriate supporting material arranged in outline form.

CONCLUSION (OF LESSON): The final portion of a period of instruction during which you re-emphasize the main points, tell students how they will apply this new learning, and close with a strong statement to leave a lasting impression of the importance of the subject learned learned.

CONFERENCE: The instructional conference is a method of instruction which develops the learning material primarily through student discussion. It includes lecture, demonstration and student performance when applicable.

DEMONSTRATION: The "showing" portion of a period of instruction; demonstrations may be followed by student performance or they may be designed as supporting material to assist student understanding.

ENTHUSIASM: An outward display which conveys the impression of intense, eager and sincere interest in the subject. You frequently reveal enthusiasm thru vigorous, confident, forceful movement, gestures and speech. However, you do not convey enthusiasm by loud, bombastic, boisterous delivery.

EVALUATION: Analysis and interpretation of the results of student proficiency measurements in terms of acceptability or the accomplishment of training objectives.

FLUENCY: A smooth flow of well-chosen words arising from clear logical thinking.

FORCE: Demonstrated vitality and strength of conviction. Aggressive decisiveness displayed thru authoritative confident manner, movement, voice and language.

GENERAL KNOWLEDGE: Level of proficiency which makes the student aware of a subject and provides certain fundamental facts and principles.

GESTURE: A natural movement of any part of the body to help convey a thought or emotion or to reinforce oral expression. A gesture does not draw attention to itself but seems a natural aid to expression.

GRAMMAR: Correct usage of spoken or written words with respect to accepted principles of number, case, and tense.

INTEREST FACTOR: Something used to gain, maintain or increase student interest in the instructional presentation. An interest factor could be a startling statement, stimulating question, skit, teaching vehicle or training aid. It is not an instructional "gimmick" to startle or entertain students but must relate to the subject matter being learned.

INTEREST SPAN: The length of time that most students in a class can give uninterrupted or continuous concentration upon the instructional presentation.

INTRODUCTION: The beginning of a period of instruction which serves to gain attention, to tell the student <u>what</u> he is to learn, to show <u>why</u> he should learn it, and to explain <u>how</u> the subject relates to past, current or future instruction and how it will be presented.

LEARNING: A change in the student's knowledge, understanding, skill or appreciation which causes him to act differently as a result. To be effective, learning must be purposeful activity which results in a change in behavior.

LECTURE: A method of instruction designed to present orally a large amount of material in a relatively short period of time with no or a minimum of student participation. Such participation is generally secured by asking brief factual questions at stated periods.

LESSON OBJECTIVE: A brief statement (in infinitive form) of the mission of the instructional period, the type of class receiving instruction and the student learning proficiency desired (general knowledge, working knowledge, qualified). The objective is the broad purpose of the lesson; the teaching points are the smaller, essential units which insure fulfillment of the objective.

LESSON OUTLINE: The systematic arrangement of the material to teach in full sentence outline form. Annex A is a sample lesson outline with Roman numerals indicating the six main portions: lesson objective, teaching points, advance assignment, introduction, body and conclusion.

LESSON PLAN: The contents of the vault file. It includes the following instructional materials: preparation data section, lesson outline section, training aids used, handouts used, official directives and comments, bibliography of references, research materials and pertinent portions of former lesson plans.

LESSON PLANNING: All the steps you follow in preparing a period of instruction from the time you receive the training directive until you actually present the lesson. These steps are divided: careful analysis of the mission, estimate of the teaching situation, preparation of lesson objective, study and research, selection of teaching points and proper supporting materials, organizing teaching points and supporting materials in a logical sequence in the written outline, deciding appropriate methods of instruction, determining good transitions and training aids, accomplishing necessary administrative details, preliminary rehearsals revisions and final rehearsals.

MANNERISM: A characteristic movement made by an individual which draws the attention of others to the movement. It normally constitutes a distraction rather than an aid to oral communication.

MEASUREMENT: The act of determining a student's knowledge, understanding, or skill by means of a device such as a performance exercise, paper and pencil test or rating form.

MOTIVATION: The process of stimulating action towards satisfying a need or reaching a goal. The creation or arousing of a desire to learn.

MURDER BOARD: A rehearsal for members of the group, who critique in detail the instructional techniques, materials, content, and estimated effectiveness for learning.

PITCH: The tone level of the voice. Your natural pitch is the midpoint of your range. This may be compared to singers who are tenor, bass, or baritone, yet each has a range of notes at his command.

PRINCIPLE: A settled rule of action or law of conduct or doctrine; a fundamental truth.

PROGRAM OF INSTRUCTION (POI): A training directive which is a combination of master training program and subject schedule. It includes course number and title, purpose, prerequisites, length, location and special annexes or blocks of instruction which in turn list the subject and problem number, length of period and type of instruction (lecture, conference, demonstration, or practical work), scope of instruction and specific study references.

QUALIFIED: Level of proficiency which includes comprehensive knowledge of subject that permits skilled functioning in a specific job and the ability to supervise or train others in that job.

QUESTIONS:

    a. LEAD-OFF QUESTION: A question (of general nature) used to start an overall discussion concerning the entire problem or broad aspects of a teaching point.
    b. FOLLOW-UP QUESTION: A question designed to narrow the area of discussion to more specific material supporting the teaching point and/or to develop ideas expressed by students responding to lead-off questions.
    c. CHECK-UP QUESTION: A question to review or summarize instruction, and to check student understanding at various points throughout the lesson.
    d. RHETORICAL: A question which is not meant to be answered by students, but by the instructor himself. It is designed to stimulate thought, to capture attention, and to lead into an area of discussion or explanation.

RAPPORT: The mutual feeling of "oneness" wherein the instructor and his students are cooperatively working toward the instructional objective.

RATE: The speaker's speed of delivery.

RESEARCH: The process of recognizing new problem situations and attempting to solve them through critical and intelligent inquiry and collection of pertinent data. This data is analyzed and evaluated in order to reach specific meaningful conclusions.

SCHOOL SOLUTION: A position, technique, or explanation which the instructor presents as the best solution for a given problem or situation based on all information and experience available at the time. The school solution is usually presented as "a school solution," implying and frequently stating that other solutions may be or are acceptable.

SCOPE OF INSTRUCTION: The problem description in the POI annexes which indicates the principal areas of that subject which will be discussed during the instructional period.

SENTENCE FORM OUTLINE: A type of lesson outline in which each main point and all supporting points are written down as complete sentences so that their meaning and relation to other points are made completely clear.

SINCERITY: An observable trait in which the speaker feels that the knowledge he possesses is valid, accurate and of great importance to others. He expresses himself naturally, simply and directly in an earnest desire to convince his listeners of the truth or value of his ideas.

SKILL: The student's ability to use his knowledge effectively. Skills may be manual or mental.

SUB-SUMMARY: A review of all or a portion of the instruction which has been presented to that point. Its purpose is to assist student understanding. Sub-Summaries may be accomplished by brief statements, by asking check-up questions, by a summary chart or projectual, by practical exercises or by having a student restate the student performance objective in his own words.

SUMMARY: That portion of the conclusion wherein the instructor reviews and reemphasizes the performance objectives of the lesson, together with the essential supporting material.

SUPPORTING MATERIAL: Any material or device which clarifies, amplifies, or reemphasizes a main idea and serves to develop maximum student understanding. Supporting material may consist of explanation, analogy, illustration, example, statistics, testimony, quotations, demonstrations, and student practice exercises.

STUDENT PERFORMANCE OBJECTIVE: A specific statement (in full sentence form) of a principle, technique, procedure or an element of essential knowledge, skill, or appreciation which students should learn as a result of a period of instruction. These points should be developed and interrelated throughout the instructional period in a logical and meaningful order.

TEACHING VEHICLE: A device which binds instruction around a central theme or story and helps to give life and movement or logical order to a presentation. Frequently, it serves as an interest factor.

TOPIC FORM OUTLINE: Similar in content to the sentence outline except the main ideas and supporting material are reduced to key words or phrases that serve to recall or suggest the complete idea.

TRANSITION: A means by which the instructor maintains logical continuity or progresses from point to point in a lesson; it is used to show relationships which assist understanding. A good method of transition is to use a teaching vehicle to which the instructor may refer from time to time to show the relative position of teaching material in context to the whole lesson. A rhetorical question or a subsummary is also a good method of transition.

VAULT FILE: The complete official reference folder relating to an instructional presentation. The problem folder includes current and former lesson plans, student handouts, descriptive copies of training aids and all data pertinent to the problem.

VISITORS FOLDER: A folder to familiarize visitors with the problem being observed. It includes problem title, length, method(s) of instruction, type of class(es) receiving instruction, lesson objective(s), the teaching points, time breakdown by teaching point, and instructional materials issued to students.

WORKING KNOWLEDGE: A level of proficiency indicating sufficient familiarity with the primary purposes, major functions, or principles of employment of a subject to permit routine practical applications.

SECTION TWO

PROGRAM OF INSTRUCTION FOR ADVANCED
PISTOL MARKSMANSHIP

CHAPTER X

TRAINING STANDARDS AND SCHEDULING COURSE OF INSTRUCTION

A. GENERAL

It must be understood that there is no substitute for knowledge and experience in teaching pistol marksmanship. An individual assigned the specific duties of teaching pistol marksmanship fundamentals, techniques of fire or coaching techniques must also be completely convinced of the importance of Advanced Pistol Marksmanship to the mission of the Armed Forces of the United States. A winning team is the pride of any commander---as it should be---whereas a losing team, while performing to the best of its ability, may possibly lose by a single point and never receive the recognition it deserves. This fact in itself necessitates a program of instruction that will increase the skill and provide the knowledge that will motivate a continuing desire to excel. Good marksmanship and good coaching has never been conveyed quickly, nor can it be gained in a short period of time. Good marksmanship is a combination of good coaching, a commanding knowledge of fundamentals, mental and manual dexterity, and a desire to excel that is willing to undergo long practice and hard work. A good instructor and coach will call on experience to aid him in his job and he will give his men the advantage of everything he has learned. With proper application of the techniques discussed in the training program, the resulting proficiency will continue producing championship teams.

B. IMPLEMENTING THE TRAINING MISSION

To properly implement the mission of training a pistol team, a comprehensive training program has been devised. The basis of this training program, of course, is the knowledge and experience of the instructor-shooters and the coaches who work with them, past, present, and future. The wealth of information on pistol shooting skill present among these personnel must be made available to other competitive shooters in the nation. The instructor-shooter must become, during his duty tour, a highly qualified instructor. The pistol coach must not only assist in training the shooter to be an instructor and a highly qualified marksman but he must also be the main tool in implementing the training program. Program organizing, instructor rehearsing, setting up of facilities, arranging for fabrication and proper use of training aids, studying refinements of various techniques and presenting instructional matter in an exemplary and interesting manner is but part of the function of a coach. To accomplish this latter mission more effectively, the instructor-coach must be familiar with advanced instructional techniques, be acquainted with the physiological-psychological processes of the human body, must learn to influence the developing shooter toward a mentally conditioned state so that he will properly respond and coordinate his thoughts and actions for control of his shooting.

C. STAGES OF DEVELOPMENT OF THE PISTOL SHOOTER

Certain factors and specific stages of development govern the progress of a pistol shooter and the instructional techniques should reflect these degrees of development and learning.

These degrees of learning are, in a sense, phases of advancement: one, exploratory; two, articulation or conversation; three, organization or resolution; four, accomplishment or consummation.

First, a comparatively new shooter usually finds that he is in a phase of exploring the skill of advanced pistol marksmanship when his interest is induced by other shooters or a coach. This period is characterized by the sensing of the important factors of a skillful endeavor. In his limited experience, he has learned that certain vital and dynamic qualities are present and his curiosity is aroused. His interest in shooting can be sustained if the one endeavoring to expand the beginner's knowledge, confines his teaching to the simpler aspects.

Second, effective instruction proceeds by making sure that exploration is followed by extensive, inquisitive conversation. A shooter's learning possibilities are limited if he continues to ponder by himself. Beyond a certain point, he must become articulate, begin to ask questions and look things up. Real conversation begins when the developing shooter does not like the answers he is getting, and offers rebuttal. A meeting of minds results and there ensues a give and take in which one position is corrected by another. Effective instruction takes place in this exchange of ideas and information between coach and shooter or shooter and shooter. Open and free communication is essential to this learning process. During this period the shooter may zig toward excessive self-doubt and zag toward excessive self-confidence but the skillful coach-instructor brings him back to the proper bearing. No longer is the shooter merely probing, again and again he has a self-justified position of logic from which he can contribute materially to the expansion of his knowledge.

Third, intelligent coaching hurdles an important challenge by leading the shooter to an organization and full grasp of all the fundamentals, principles and factors that contribute to the controlling of his skill as a marksman. The unanswered questions are being answered satisfactorily, the problems are being resolved. There comes a time when the technique is established. This phase of organization or resolution is marked by the shooter's deciding upon an acceptable method of approach. It is also a characteristic of this period that the shooter has evolved a system of operation peculiar to him and figuratively speaking, he strikes out on his own. There has come into being a technique for accomplishment.

Fourth and last, the phase of consummation. Teaching becomes a sustaining force when the inducement to accomplishment is reached. Coaching and instructing is now counseling and advisement, a keeping on the track, so to speak. The shooter has assimilated knowledge, accepts a method of operation, exercises skillful judgement, can control vagaries of the will and has correlated various courses of action into a coefficient of the kind of shooter he now is, how he will accomplish his end. After this he emerges as a master of pistol competition. His degree of accomplishment depends upon the peculiar nature of this entity. The limitations if any, lie in the ambition, force of character, completeness of knowledge, enthusiasm and being associated with group incentive. Whether he will eventually fulfill his desire to become a champion is ultimately only his business.

D. <u>SCHEDULING COURSES OF INSTRUCTION</u>

There follows an example of a five day period of instruction on pistol marksmanship. This is to be utilized mainly for abbreviated courses of instruction with a minimum of practical work, this is, range practice and coach-shooter team work. The ideal period of time on which to base this cycle of training is approximately four (4) weeks. The expanded schedule would include an increase in the hours of practical work and repetition of certain basic phases of instruction that need reiteration. An example of a weekly training schedule is found in Chapter 12 "Pistol Team Organization and Administration".

ADVANCED PISTOL MARKSMANSHIP INSTRUCTOR TRAINING PROGRAM
(5 DAY TRAINING SCHEDULE)

1ST DAY

TIME	SUBJECT	INSTRUCTOR	REMARKS
0750-0800	Introduction	Pistol Div OIC	
0800-0825	Team Org. and Administration	Instr-Shooter	
0830-0855	Procurement, maintenance and security of weapons.	Instr-Shooter	
0900-1050	Match Rules & Range Safety	Instr-Shooter	
1100-1150	Fundamentals I (Attaining a Minimum Arc of Movement)	Instr-Shooter	

1ST DAY (CONT)

TIME	SUBJECT	INSTRUCTOR	REMARKS
1330-1420	Fundamentals II (Sight Alighment)	Instr-Shooter	
1430-1520	Fundamentals III (Trigger Control)	Instr-Shooter	
1530-1630	Zeroing .22 Cal & .45 Cal (Service) 25 Yds	Instr-Shooter	10 rds ea, caliber

2ND DAY

TIME	SUBJECT	INSTRUCTOR	REMARKS
0800-0850	Technique of Fire I (Establishing a System)	Instr-Shooter	
0900-0950	Technique of Fire II (Slow Fire)	Instr-Shooter	
1000-1200	Slow Fire Practice - 50 rds	Instr-Shooter	10 rds ea .22 cal & .45 cal
1330-1420	Technique of Fire III (Sustained Fire)	Instr-Shooter	
1430-1500	Sustained Fire Drill - 25 yds	Instr-Shooter	20 rds ea .22 cal & .45 cal
1500-1630	Timed & Rapid Fire Practice 25 yds	Instr-Shooter	20 rds ea. .22 cal & .45 cal

3RD DAY

TIME	SUBJECT	INSTRUCTOR	REMARKS
0800-0850	Attributes, Responsibilities and Duties of a Pistol Team Coach	Instr-Shooter	
0900-1050	Technique of Coaching a Pistol Team	Instr-Shooter	
1100-1125	Evaluation of the Team Shooter	Instr-Shooter	
1330-1500	.22 Cal NMC - 50 & 25 yds	Instr-Shooter	30 rds .22 cal
1500-1630	.45 Cal (Service) NMC 50 & 25 yds	Instr-Shooter	30 rds .45 cal

4TH DAY

TIME	SUBJECT	INSTRUCTOR	REMARKS
0800-0850	Mental Discipline	Instr-Shooter	
0900-0950	Effect of Tobacco, Alcohol, Coffee and Drugs	Instr-Shooter	
1000-1025	Physical Conditioning & Diet	Instr-Shooter	
1035-1100	Review of Course and Weak Points	Instr-Shooter	
1330-1500	.22 Cal NMC Team Match 50 & 25 yds	Instr-Shooter	30 rds .22 cal
1500-1630	.45 Cal (Service) NMC Team Match - 50 & 25 yds	Instr-Shooter	30 rds .45 cal

5TH DAY

TIME	SUBJECT	INSTRUCTOR	REMARKS
0800-1000	.22 NMC Team Match - 50 & 25 yds	Instr-Shooter	30 rds .22 cal
1000-1200	.45 Cal (Service) NMC Team Match - 50 & 25 yds	Instr-Shooter	30 rds .45 cal
1330-1530	Written Examination, Critique and Panel Discussion	Instr-Shooter	
1530-1630	Award of Certificates	Pistol Div, OIC	

E. TRAINING AIDS, INSTRUCTIONAL MATERIAL AND SUPPLIES NEEDED TO CONDUCT AN INSTRUCTOR COURSE IN PISTOL MARKSMANSHIP TRAINING.

    1. Slide Projector with screen.

    2. Set of 35mm slides.

    3. Blackboard, chalk with easel and eraser.

    4. Podium cards and podium.

    5. Lesson plan and notes.

    6. Watch.

    7. Pointer.

    8. GTA charts if not using film slides. (Complete list of GTA charts for Advanced Pistol Marksmanship Course is prescribed in annex on following pages, this chapter).

    9. Public address system.

    10. Pertinent Army regulations and directives.

    11. Three (3) Pistol boxes with complete equipment.

    12. Classroom or bleachers.

    13. NRA Pistol Rule Book (Current Edition).

    14. Advance Pistol Marksmanship Manual (One copy each student).

    15. Sight adjustment cards (one copy each student).

    16. Pencil and paper (Each student).

    17. NRA Scorecards (Single stage, Multistage and Team Matches).

    18. Practice Scorecards (AMTU Form 8).

    19. Training schedule of Instructor Course (Each student).

    20. Certificates of successful completion (Each student).

    21. Sling shot device to demonstrate sight alignment error.

    22. Slow Fire Worksheet (Each student).

    23. Timed and Rapid Fire Worksheet (Each student).

    24. Line Coach Worksheet (Each student).

    25. Students comment sheet (Each student).

26. Information questionnaire (Each student).

27. Examination (Each student).

28. Roster of students.

29. Master score sheet with coach, team and target assignments.

30. Ammunition requirements for each day of course.

31. Masking tape and chart stripping material.

32. Hammer and nails.

33. Scissors.

34. Magic markers (Assorted colors).

35. Straightedge (Graduated scale).

36. Scratch pads.

37. Engraved trophy for high non-distinguished student in .45 cal HB Team Match.

38. A .22 Cal and a .45 Cal Service ammunition pistol for each student. (Students should be directed to procure necessary weapons at parent unit.)

39. Scorebooks (Each student).

40. Grease marking pencils (Assorted colors).

F. <u>CERTIFICATE OF TRAINING.</u> (Dept of Army Form 87.)

Department of the Army

Certificate of Training

This is to certify that

SFC RICHARD ROE

has successfully completed
THE PISTOL MARKSMANSHIP INSTRUCTOR COURSE
CONDUCTED BY THE
UNITED STATES ARMY MARKSMANSHIP TRAINING UNIT

Given at FORT BENNING, GEORGIA - 11 JANUARY 1964

JOSEPH E DOE
Major, Infantry
OIC, Pistol Division

DA FORM 87
1 SEP 64   REPLACES DA FORM 87, 1 JAN 49, WHICH IS OBSOLETE

ANNEX TO TRAINING STANDARDS AND SCHEDULING COURSE OF INSTRUCTION (I-10)

GRAPHIC TRAINING AIDS REQUIREMENT FOR CONDUCTING THE
ADVANCED PISTOL MARKSMANSHIP TRAINING COURSE

CHART # I

    <u>FUNDAMENTALS I</u>   (ATTAINING A MINIMUM ARC OF MOVEMENT)

        1. STANCE

        2. POSITION

        3. GRIP

        4. BREATH CONTROL

CHART # II

    <u>FUNDAMENTALS II</u>   (SIGHT ALIGNMENT)

        1. RELATIONSHIP OF SIGHTS

        2. POINT OF FOCUS

        3. CONCENTRATION

        4. CHARACTERISTICS OF THE HUMAN EYE

CHART # III

    <u>FUNDAMENTALS III</u>   (TRIGGER CONTROL)

        1. NERVE PROCESSES

        2. FACTORS PROVIDING FOR THE CORRECT CONTROL OF THE TRIGGER

        3. APPLICATION OF POSITIVE TRIGGER PRESSURE

CHART # IV

    <u>TECHNIQUE OF FIRE I</u>   (ESTABLISHING A SYSTEM)

        1. PREPARATION FOR SHOOTING

        2. PLAN SHOT SEQUENCE

        3. RELAXATION BEFORE SHOT OR STRING

        4. DELIVER SHOT OR STRING OF SHOTS ON TARGET

        5. MAKE AN ANALYSIS OF EACH SHOT OR STRING

        6. MAKE A POSITIVE CORRECTION IF NECESSARY

CHART # V

TECHNIQUE OF FIRE II (SLOW FIRE)

1. EMPLOYMENT OF THE FUNDAMENTALS
2. TECHNIQUES IN SLOW FIRE CONTROL
3. COMMON DEFICIENCIES IN CONTROL
4. WIND SHOOTING AND ADVERSE CONDITIONS
5. TRAINING METHODS

CHARTS # VI

TECHNIQUES OF FIRE III (SUSTAINED FIRE)

1. EMPLOYMENT OF THE FUNDAMENTALS
2. TECHNIQUE OF SUSTAINED FIRE
3. COMMON DEFICIENCIES IN CONTROL
4. TRAINING METHODS
5. WIND SHOOTING AND ADVERSE CONDITIONS

CHARTS # VII

MENTAL DISCIPLINE

1. ESSENTIAL TO ADVANCED PISTOL MARKSMANSHIP
2. DEVELOPING MENTAL DISCIPLINE AND CONFIDENCE
3. WHY CAN'T YOU BE A WINNER?
4. MATCH PRESSURE
5. REDUCING TENSION AND ATTAINING RELAXATION

CHART # VIII

COACHING I (ATTRIBUTES, RESPONSIBILITIES AND DUTIES OF A PISTOL TEAM COACH)

1. PERSONAL ATTRIBUTES OF A COACH
2. HEAD COACH RESPONSIBILITIES
3. LINE COACH DUTIES

CHARTS # IX

    COACHING II   (TECHNIQUE OF COACHING A PISTOL TEAM)

        1. COACHING THE CHAMPIONS

        2. FACTORS IN SYSTEMATIC COACHING TECHNIQUE

        3. HEAD COACH TECHNIQUE

        4. LINE COACH TECHNIQUE

        5. SUGGESTION FOR IMPROVING COACHING TECHNIQUE

        6. DEMONSTRATION FOR COACHING TECHNIQUE

CHART # X

    COACHING III   (EVALUATION OF THE PISTOL TEAM SHOOTER)

        1. INDIVIDUAL INFORMATION AND EVALUATION SHEET

            a. LINE COACH EVALUATION

            b. PROGRESS GRAPH

            c. AGGREGATE RECORD

        2. EXAMINATION

        3. SCORE BOOK

        4. ATTRIBUTES OF A TEAM SHOOTER

CHART # XI

    PHYSICAL FITNESS I   (PHYSICAL CONDITIONING)

        1. BASIS FOR A GOOD PHYSICAL CONDITION

        2. TYPES OF EXERCISE

        3. PISTOL TEAM DAILY DOZEN EXERCISES

        4. STATIC TENSION EXERCISES

CHART # XII

    PHYSICAL FITNESS II   (DIET AND HEALTH OF THE COMPETITIVE SHOOTER)

        1. GENERAL INFORMATION

        2. IMPORTANCE OF PROPER DIET AND NUTRITION

        3. DIET AND CARE OF HEALTH

CHART # XIII

    <u>PHYSICAL FITNESS III</u>   (EFFECTS OF ALCOHOL, COFFEE, TOBACCO AND DRUGS)

        1. ALCOHOL

        2. COFFEE AND TEA

        3. TOBACCO

        4. DRUGS

CHART # XIV

    <u>COMPETITIVE REGULATION I</u>   (NRA PISTOL MATCH RULES)

        1. TYPES OF COMPETITION

        2. COMPETITION REGULATIONS

        3. CHALLENGES AND PROTESTS

        4. MATCH WEAPONS, EQUIPMENT AND AMMUNITION

        5. SCORING

        6. COMPETITOR'S DUTIES AND RESPONSIBILITIES

        7. TEAM OFFICERS' DUTIES AND POSITION

        8. ELIGIBILITY OF COMPETITORS

CHART # XV

    <u>COMPETITIVE REGULATIONS II</u>   (NRA PISTOL RANGE PROCEDURE AND SAFETY RULES)

        1. RANGE CONTROL

        2. RANGE FIRE COMMANDS

        3. PISTOL RANGE SAFETY RULES

CHART # XVI

    <u>COMPETITIVE REGULATIONS III</u>   (NATIONAL TROPHY PISTOL MATCH RULES AND REQUIREMENTS FOR A DISTINGUISHED PISTOL SHOT BADGE)

        1. NATIONAL TROPHY PISTOL MATCH CONDITIONS (AR 920-30 CONDENSED)

        2. DISTINGUISHED PISTOL SHOT BADGE.

CHART # XVII

> ADMINISTRATION I   (TEAM ORGANIZATION AND ADMINISTRATION)
>
>> 1. TEAM ORGANIZATION
>>
>> 2. ADMINISTRATIVE RESPONSIBILITY
>>
>> 3. ADMINISTRATIVE REQUIREMENTS

CHART # XVIII

> ADMINISTRATION II   (PROCUREMENT, MAINTENANCE AND SECURITY OF WEAPONS EQUIPMENT AND AMMUNITION)
>
>> 1. SECURITY
>>
>> 2. PROCUREMENT
>>
>> 3. CAUSES OF WEAPONS MALFUNCTIONS AND CARE AND MAINTENANCE
>>
>> 4. THE MARKSMANSHIP UNIT SHOP IS PART OF THE TEAM EFFORT

CHART # XIX

> REVIEW OF COURSE
>
>> 1. FUNDAMENTALS I   (ATTAINING A MINIMUM ARC OF MOVEMENT)
>>
>> 2. FUNDAMENTALS II   (SIGHT ALIGNMENT)
>>
>> 3. FUNDAMENTALS III   (TRIGGER CONTROL)
>>
>> 4. TECHNIQUE OF FIRE I   (ESTABLISH A SYSTEM)
>>
>> 5. TECHNIQUE OF FIRE II   (SLOW FIRE)
>>
>> 6. TECHNIQUE OF FIRE III   (SUSTAINED FIRE)
>>
>> 7. MENTAL DISCIPLINE
>>
>> 8. COACHING I   (ATTRIBUTES, RESPONSIBILITIES AND DUTIES OF A PISTOL TEAM COACH)
>>
>> 9. COACHING II   (TECHNIQUE OF COACHING A PISTOL TEAM)
>>
>> 10. COACHING III (EVALUATION OF THE PISTOL TEAM SHOOTER)

CHART # XX

> REVIEW OF COURSE (Cont'd)
>
>> 11. PHYSICAL FITNESS I (PHYSICAL CONDITIONING)

12. PHYSICAL FITNESS II (DIET AND HEALTH OF THE COMPETITIVE SHOOTER)

13. PHYSICAL FITNESS III (EFFECTS OF ALCOHOL, COFFEE, TOBACCO AND DRUGS)

14. COMPETITIVE REGULATIONS I (NRA PISTOL MATCH RULES)

15. COMPETITIVE REGULATIONS II (NRA PISTOL RANGE PROCEDURE AND SAFETY RULES)

16. COMPETITIVE REGULATIONS III (NATIONAL TROPHY PISTOL MATCH RULES AND REQUIREMENTS FOR A DISTINGUISHED PISTOL SHOT BADGE)

17. ADMINISTRATION I (TEAM ORGANIZATION AND ADMINISTRATION)

18. ADMINISTRATION II (PROCUREMENT, MAINTENANCE AND SECURITY OF WEAPONS, EQUIPMENT AND AMMUNITION)

19. REVIEW OF WEAK POINTS

ANNEX TO TRAINING STANDARDS AND SCHEDULING COURSE OF INSTRUCTION (I-10)

STUDENT COMMENTS, ADVANCED PISTOL MARKSMANSHIP INSTRUCTORS COURSE

    1. Did you receive a proper orientation and instruction concerning this course prior to attending this class? If so by whom?

    2. Which period of instruction in this course gave you the most benefit? Explain.

    3. Which period(s) of instruction in the course do you feel can be improved? Tell how.

    4. Do you have any suggestion(s) in training better coaches and shooters which this course failed to bring out? If so, explain briefly.

    5. Do you feel that you have an improved understanding of pistol coaching and shooting from this course? How?

    6. What are your comments about the course in general? Write them; continue on the reverse side if this space is insufficient.

    7. Please turn this comment sheet into the coach-shooter-instructors present.

_____ _____
  (RANK)   (NAME)   (SERIAL NUMBER)   (ORG.)

PLEASE PRINT

CHAPTER XI

LESSON PLANS FOR THE ADVANCED PISTOL MARKSMANSHIP COURSE

FUNDAMENTALS I (P-1)

ATTAINING A MINIMUM ARC OF MOVEMENT

LESSON OUTLINE

I. LESSON OBJECTIVE: To impart to coaches and instructor-shooters an understanding of the fundamental factors involved in attaining a minimum arc of movement and why they are deemed essential to the development of a system of maximum control of their shooting skills.

II. STUDENT PERFORMANCE OBJECTIVES: As a result of this instruction, the student must be able to:

A. EXPLAIN the main requirements of a proper stance, understand on a general basis the physical characteristics of the human body that affect attaining a proper stance and be able to devise and assume the most advantageous stance that meets his personal needs.

B. EXPLAIN the requirement of a proper position, understand how to orient the body for assuming a proper position and why uniformity and naturalness are essential to a perfect position.

C. EXPLAIN the main requirements of a proper grip, understand how to attain and check out a proper grip and know the various methods of developing and strengthening the grip of the shooting hand.

D. EXPLAIN the main requirements of proper breath control, understand the physiological processes of breathing and demonstrate the proper method of breath control prior to and in conjunction with fire commands.

III. ADVANCE ASSIGNMENT: The Advanced Pistol Marksmanship Manual Section I "Fundamentals I of Advanced Pistol Marksmanship". Chapter I, "Attaining a Minimum Arc of Movement."

IV. INTRODUCTION: (2 Min)

A. Gain Attention: Have your pistol scores improved enough in the last five years to enable you to win a pistol match?

B. Orient Students:

1. Lesson Tie-in: Advanced pistol marksmanship is the highly skilled and trained part of competitive pistol shooting that enables the informed master pistol shooter to fire winning scores. During this hour we shall discuss in great detail how the fundamentals that comprise attaining minimum arc of movement are the foundation of advanced pistol marksmanship.

2. Motivation: Champions are not born. They are the ultimate evolution of a development process that learns well all the fundamentals and thru intense, continuous hard work gain the ability to control the employment of the fundamentals in practice and in matches. Good scores do not come easy. Concentration on learning the fundamentals is the important consideration in readying oneself to master the pistol.

NOTES: USE PYRAMID OF MARKSMANSHIP EXAMPLE.

3. <u>Scope</u>: During this period we will discuss the first four basic fundamentals of pistol marksmanship. You will be able to attain a satisfactory minimum arc of movement by learning proper stance, position, grip and breath control.

V. BODY: (8 Min)

   A. <u>First Student Performance Objective</u>: The shooter must be able to explain the main requirements of proper stance, be familiar with the physical characteristics of those parts of the human body employed in attaining proper stance and devise and assume the most advantageous stance to meet his personal needs.

      <u>QUESTION</u>: What factors make up a good stance?

      1. Main requirements of stance.

         a. Equilibrium and stability without strain on muscles.

         b. Greatest degree of immobility - absence of any independent movement of any part of the body.

         c. Head position for best use of eyes.

<u>NOTE</u>: TURN CHART, USE DEMONSTRATOR.

      2. Characteristics of the human body that affect stance.

         a. Passive apparatus (Bones and Ligaments)

         b. Active apparatus (Muscles)

         c. Nervous system

         d. Vestibular system (Balance or Equilibrium)

      3. Assuming the Stance.

         a. Separation of feet

         b. Stand erect

         c. Legs straight

         d. Hips level, abdomen relaxed

         e. Head and shoulders level

         f. Non-shooting arm and hand relaxed

         g. Body balanced slightly forward

TRANSITION: It is insufficient to merely assume a comfortable and stable stance. You should be able to point the shooting arm naturally at the target.

B. Second Student Performance Objective: (8 Min)
The shooter must be able to explain the main requirements of proper position, understand how to orient the body for assuming a proper position and know why uniformity and naturalness are essential to a perfect position.

QUESTION: Why does the shooter have to be so careful of getting a proper position? Why can't he just point at the target and shoot?

NOTE: UNCOVER GTA ON "POSITION" USE DEMONSTRATOR.

1. Main requirement of a proper position:

   a. The arm and body should point naturally at center of target.

2. Method of Orientation:

   a. Face to left of target 40 to 50 degrees.

   b. Turn head to face directly toward target.

   c. Raise shooting arm to target and close eyes.

   d. Swing the body as a unit, pivoting from the ankles, stop and settle into a natural point.

   e. Open eyes and check alignment of arm to target.

   f. If it deviates from center, move rear foot in direction of error.

TRANSITION: Good stance and position will give you a good platform to shoot from. You also need a firm grip on the pistol that will permit the sights to fall naturally into alignment when the arm is extended for the delivery of a shot.

C. Third Student Performance Objective: The shooter should be able to explain the main requirements of a proper grip, understand how to attain and check out a proper grip and know the various methods of developing and strengthening the shooting hand.

QUESTION: What is the most important feature of a proper grip?

NOTE: UNCOVER GTA ON "GRIP". USE DEMONSTRATOR w/.45 PISTOL.

1. Uniformity.

2. Requirements of a good grip.

   a. Natural sight alignment.

   b. Grip firmly to prevent shifting of grip.

   c. No change in tightness during firing.

   d. Independent movement of trigger finger.

   e. No change in character of grip.

    f. Grip must be comfortable.

    g. Force of recoil straight to rear into arm and shoulder.

    h. Do not grip for long periods. Prevent undue fatigue of shooting hand.

3. Method of getting proper grip.

    a. Hold pistol by barrel of non-shooting hand.

    b. Spread index finger and thumb to form a "V".

    c. Bend wrist downward slightly.

    d. Seat the pistol into the "V" of the shooting hand.

    e. Fit pistol firmly into gripping space between thumb and base of index finger

    f. Place trigger finger above or beside the trigger guard.

    g. Grasp stock with lower three fingers.

    h. Thumb placed high on left side of stock.

    i. Vise-like action of primary pressure points.

    j. Place trigger finger on trigger so as to allow straight back pressure on trigger.

    k. Tighten grip to maximum without tremor.

4. Checking for proper grip.

    a. Natural sight alignment.

    b. Grip firm enough to prevent shifting of grip.

    c. Variations of tightness or character of grip.

    d. Independent trigger action.

    e. Comfortable grip.

    f. Force of Recoil straight to rear into arm and shoulder.

5. Aids to developing a good grip.

    a. Use exerciser for building strong grip (rubber ball, gripping spring, etc.)

    b. Never casually change your grip. Careful analysis that dictates a modification that will insure improvement is the only justification.

    c. Adapt grip to different types and calibers of guns.

d. Shaped, molded or custom grips.

e. Rosin can help under some circumstances.

TRANSITION: Good stance, position and grip will give the shooter stability, balance and solid arm support but not immobility. A shooter may have excess body movement if he can't control his breathing during the delivery of a shot.

    D. Fourth Student Performance Objective: (4 Min)

The shooter must be able to explain the requirements of proper breath control, understand the physiological processes of breathing and demonstrate the proper methods of breath control prior to and in conjunction with fire commands.

QUESTION: Why is control of breath so important?

NOTE: UNCOVER CHART ON "BREATH CONTROL." USE DEMONSTRATOR.

1. Employed systematically and Uniformly.

    a. Promotes a steady hold.

    b. Utilize the physiological process of breathing.

    c. Comfortable and relaxed.

2. Recommended Method.

    a. Prior to fire commands.

    b. During firing commands.

NOTE: ASK FOR AND ANSWER STUDENT QUESTIONS.

  VI. CONCLUSION:

    A. Retain Attention: The minimum arc of movement (the steadiest, stillest hold you are capable of) is your foundation for shooting.

    B. Summary: Key points of:

NOTE: UNCOVER GTA, DISMISS DEMONSTRATOR.

1. Stance - stable, balanced, immobile with good head position.

2. Position - Naturally pointing at target.

3. Grip - Must be uniform, the requirements, how to check out grip and how to develop a strong grip.

4. Breath Control - Must be Systematic and Uniform, the main requirements, use the physiological processes and how to coordinate with fire commands.

    C. Application: These fundamentals are the foundation upon which to build good scores. Learn them completely and put them into practice.

    D. <u>Closing Statement</u>: In all practice sessions, as well as in matches, the shooter must concentrate on the mastering of the minimum arc of movement. Only by so doing can we expect to show a constant improvement and become a consistent, better shooter.

<u>PREPARATION DATA SECTION</u>

I.	TITLE AND NUMBER:	Fundamentals I: Attaining a minimum arc of movement (P-1)
II.	TIME ALLOTTED:	Fifty (50) minutes
III.	PERIODS OF INSTRUCTION:	One
IV.	CLASS PRESENTED TO:	Coaches and Instructor Shooters
V.	INSTRUCTOR REFERENCES:	The Advanced Pistol Marksmanship Manual
VI.	INSTRUCTIONAL AIDS:	2 copies of lesson plan, hand notes, 1 podium 1 sound set, GTA charts, or 35mm slides & projector, podium cards, blackboard, chalk, eraser, pointer, 1 table, 1.45 cal pistol.
VII.	STUDENT EQUIPMENT & UNIFORM:	Uniform as prescribed
VIII.	PHYSICAL FACILITIES:	Classroom or outdoor bleachers for students.
IX.	PERSONNEL REQUIREMENTS:	1 principal instructor, demonstrators and assistant instructors as needed, 1 sound & projector operator.

FUNDAMENTALS II (P-2)

SIGHT ALIGNMENT

I. LESSON OBJECTIVE: To enable coaches and instructor-shooters to understand, teach and apply the factors necessary to a system of maximum control of their shooting.

II. STUDENT PERFORMANCE OBJECTIVES: As a result of this instruction the student must be able to:

    A. EXPLAIN the relationship of the front and rear sights as it pertains to attaining perfect sight alignment and understand how to apply it in controlling the strike of the bullet on the target.

    B. DEFINE the point of focus and the reason for it.

    C. EXPLAIN the duration of intense, uninterrupted mental concentration and how anxiety affects the sight relationship.

    D. DISCLOSE the limiting characteristics and optical properties of the human eye.

III. ADVANCE ASSIGNMENT: The Advanced Pistol Marksmanship Manual, Chapter II, "Sight Alignment".

IV. INTRODUCTION:

    A. <u>Gain Attention</u>: Bill Blankenship, many times the National Pistol Champion, has won most of the pistol records and titles in the United States. All of those shooters who have held or presently hold records and titles did not gain this enviable stature by black magic. They received such acclamation because they knew how to control the application of the fundamentals to their shooting.

    B. <u>Orient Students</u>:

        1. <u>Lesson Tie-In</u>: In previous classes you have discussed many of the fundamentals such as position, stance, breath control and grip. Another fundamental, and considered by many champions to be the most important fundamental in shooting, is that of sight alignment.

        2. <u>Motivation</u>: To understand the fundamentals is to solve your shooting problems. That is, to understand what they are and how to apply them to yourself. You must have a full working knowledge of sight alignment before you can develop into a good pistol shot. A good shooter, however, knows from experience that all the fundamentals must be mastered if he is to become a major competitor.

        3. <u>Scope</u>: You will learn to master sight alignment by knowing:

            a. How to explain relationship of the front and rear sights as it pertains to attaining perfect alignment and how to apply it in controlling the strike of the bullet on the target.

            b. How to define point of focus and the reason for it.

            c. The duration of intense, uninterrupted mental concentration and the effect anxiety has on the sight relationship.

    d. How to use your eyes to the best advantage for shooting commensurate with the limiting characteristics and optical properties of the human eye.

  V. BODY:

    A. <u>First Student Performance Objective</u>: Students must be able to explain the relationship of the front and rear sights as it pertains to attaining perfect sight alignment, and understand how to apply this factor in controlling the strike of the bullet on the target.   (12 min)

   <u>QUESTION</u>: What is the definition of perfect sight alignment?

    1. Perfect sight alignment is the proper relationship of the front sight to the rear sight as viewed by the shooter's eye that gives a definite direction of the bore of the pistol.

    2. The essence of accurate shooting lies in accurately hitting the center of a target which is comparatively small in dimension.

    3. In order for the bullet to hit the center of the target, it is necessary for the shooter to aim the barrel of the pistol in a definite direction relative to the target.

    4. In principle, accurate aiming derives from the shooter placing in exact alignment, the front sight and the rear sight, holding this alignment in his or her aiming area and thereby giving the pistol the proper direction in relation to the target.

    5. We must consider three factors for proper understanding of sight alignment. First we must have something at which to aim. This can be the target that we use in competition.

<u>NOTE</u>: SHOW CHART ONE WHICH IS A 50 YARD TARGET ON A CLEAR TRANSPARENT
    SHEET.

    6. The target is a stationary element and we will think of the bullseye in the center as our reference point only.

<u>NOTE</u>: SHOW CHART TWO WHICH IS A DRAWING OF THE REAR SIGHT ON A CLEAR TRANS
    PARENT SHEET AND FIT IT IN PERFECT ORDER UNDER THE BLACK OF THE 50
    YARD TARGET AS IN A SIX O'CLOCK HOLD.

    7. The second factor that we must consider: The rear sight. It is necessary to control the rear sight because it contains the notch where-in the front sight is to be centered.

<u>NOTE</u>: SHOW CHART THREE WHICH IS A DRAWING OF A FRONT SIGHT ON A CLEAR
    TRANSPARENT SHEET. FIT IT IN PERFECT ORDER UNDER THE BLACK CENTER
    OF THE 50 YARD TARGET. POSITION THE FRONT SIGHT IN A SIX O'CLOCK HOLD
    AND CENTER IT IN THE MIDDLE OF THE REAR SIGHT NOTCH WITH THE TOP OF
    THE FRONT SIGHT LEVEL WITH THE TOP OF THE REAR SIGHT.

    8. The third factor that must be considered is the front sight. It is centered in the rear sight notch with an equal amount of light on either side of the front sight, the top of the front sight is level with the top of the rear sight.

    9. What you see here is perfect sight alignment superimposed on a target that is in clear focus. This is referred to as perfect sight picture. It is a physical impossibility for the eye to transmit this picture to the mind when actually shooting, so forget it.

10. Before showing you what you should see in proper sight alignment, there are some facts you must understand.

NOTE: SHOW CHART FOUR WHICH IS A DRAWING OF ANGULAR SHIFT ERROR AS FOUND ON PAGE 40 OF THE PISTOL MARKSMANSHIP MANUAL.

11. If you do not keep proper alignment of your sights you will have a deviation in the strike of your bullet on the target.

NOTE: SHOW CHART FIVE WHICH IS A 50 YARD TARGET. THE SIZE DEPENDS ON THE SIZE SLINGSHOT USED. (SEE S-3 USAMTU FOR SLINGSHOT DEVICE.)

12. When the front sight moves to the right it will move the strike of the bullet to the right as shown by the dot on the target.

NOTE: PUT SLINGSHOT DEVICE AWAY AND COVER OR REMOVE CHART FIVE.

TRANSITION: You must realize the importance of sight alignment and what happens when the sights get out of alignment. Remember, you must forget the perfect sight picture as previously illustrated; it is physically impossible to see this picture when aiming a weapon at the target. There is a practical and logical answer to what we must properly see when aiming at the target.

B. Second Student Performance Objective: Students must be able to define the point of focus and the reason for it. (12 min)

QUESTION: What do we see in proper sight alignment?

1. We will use the process of elimination to answer that.

2. Correct sight alignment is the relationship of the front and rear sight only.

3. Remember that the target is stationary and since it does not move in depth or horizontally, there is no need to focus the vision on it. The target may appear to move in relation to the extended shooting arm but this illusion is caused by your minimum arc of movement. This fact eliminates the target as a point of focus.

NOTE: SHOW CHART SIX WHICH IS DRAWN ON A CLEAR TRANSPARENT SHEET AND SHOWS A 50 YARD TARGET THAT IS HAZY OR FUZZY.

4. The rear sight is over the pivot point of the wrist and is relatively stable. Eliminate the rear sight as a point of focus.

NOTE: SHOW CHART SEVEN WHICH IS DRAWN ON A CLEAR TRANSPARENT SHEET AND SHOWS A SLIGHTLY FUZZY OR ALMOST CLEARLY DEFINED REAR SIGHT. IT WILL FIT ON CHART SIX AS CHART TWO FITS ON CHART ONE.

5. That leaves the last and most important factor, the front sight. This element alone must be our point of focus. The target must remain out of focus if the shooter is to attain perfect sight alignment. The relationship of the front sight to the rear sight is the primary consideration and to accomplish a precise relationship, they must be seen distinctly.

NOTE: SHOW CHART EIGHT WHICH IS A CLEAR DISTINCT DRAWING OF THE FRONT SIGHT ON A CLEAR TRANSPARENT SHEET AND FITS ON CHART SEVEN. CHART SIX, WITH THE HAZY BULLSEYE, IS THE BACKGROUND.

6. The front sight has a tendency to move in the rear sight notch. For this reason, constant attention must be given to it. For this concentrated attention to result in perfect sight alignment, the position of the front sight must be clearly defined in relation to the rear sight notch.

7. Front sight point of focus is imperative because continuous, intense concentration is necessary in maintaining sight alignment. The degree of our control in the delivery of a good shot is reduced in proportion to the lessened degree of concentration on maintaining the correct relationship of front and rear sight.

8. Misplacing the eye focus forward to the target, or somewhere between the front sight and the target, magnifies the movement of the shooting arm and renders the front and rear sight indistinct. Proper relationship of front and rear sight cannot be maintained.

9. The front sight must be clear and distinct, the rear sight is clear or it may be slightly hazy or fuzzy, yet easily discernable. The target is hazy or fuzzy and rather indistinct in outline.

TRANSITION: In order to maintain good sight alignment you must focus the vision on the front sight, keeping it centered in the rear sight notch with the top of the rear sight and top of the front sight level. Keeping our point of focus on the front sight requires a tremendous amount of effort and concentration especially if you are in the habit of flicking your focus back and forth rapidly and often between front sight and target.

C. Third Student Performance Objective: Students must be able to explain the duration of intense, uninterrupted, mental concentration and how anxiety affects the sight alignment relationship.

QUESTION: How long can the average shooter concentrate on sight alignment without interruption?

1. A genius can concentrate on only one thing to the exclusion of all else for about eight to ten seconds. The three to six seconds the average pistol shooter can concentrate without interruption, dictates that he must fire promptly or desist and bench the weapon. Replan, relax and make a new effort to fire the shot.

2. If the sights are not fully and correctly aligned because of inability or carelessness, obviously they serve to only guide the bullet to some point other than the one desired.

3. Random thoughts of anxiety about the apparently stationary pressure on the trigger will generate an impulse to get more pressure on the trigger so it will release the hammer. This dilution of thought generally results in a momentary lessening of the concentration on sight alignment. The maintenance of perfect sight alignment suffers and the added trigger pressure usually fires the pistol before concentration returns to the front sight.

TRANSITION: Knowing the critical nature of sight alignment, we must use every element within our grasp to insure that we perform to the utmost in applying this important fundamental. Of the many faculties you and I have to help us in this demanding task, the eye is one of the most essential.

D. Fourth Student Performance Objective: Students must be able to understand the limiting characteristics and optical properties of the human eye. (9 min)

1. Optical properties of the human eye.

    a. As is well known, the aiming process makes very exacting demands upon the vision. Consistency and degree of accuracy are directly dependent upon the sharpness of vision and the conditions determining it.

        (1) The eye function is similiar to the operation of a camera. We see objects because of the light rays reflecting from them are relayed to the sensory portion of the brain by nerves leading from the retina of the eye.

        (2) Recall the description of a perfect sight picture. The reason that it is physically impossible to see this picture exactly is that the eye is constructed in such a way that it is not able to see sharply and simultaneously objects located at varying distances from it.

        (3) Rapid changing of the focus of the eye back and forth to objects at varying distances tend to tire the eye muscles, causing fatigue.

    b. All shooters should make a conscious effort to improve the condition of their eyes in the interval when they are not actually aiming by allowing the habits of normal sight to function.

        (1) Blinking is the first habit of normal sight and is an involuntary action. Lubrication, cleansing and momentary rest of the eye is the function here.

        (2) Central fixation is to have the eye and the mind so coordinated that they fix on a small area at one and the same time.

        (3) The third habit of normal eyes is shifting. Shifting the gaze prevents staring which is the commonest form of eye strain.

        (4) The fourth habit is the eye's ability to adjust for various light intensities

2. As a result of optical imperfections of the human eye, there exists a certain limit of varying sensitivity of our eyes which determines the sharpness of vision. If we go beyond these limitations or if vision is defective, these inherent imperfections have a telling effect on accuracy of shooting.

    a. Spherical abberation is the condition where the lens of the eye does not focus the rays of light to a single point on the retina of the eye.

    b. Diffraction of light is the diffusion of light rays after they enter the eye. Because of passing through the lens and fluid (humor) of the eye that is not perfectly transparent the rays are dissipated. Hence the halo effect when one views a brightly illuminated object against a dark background.

    c. Brilliant sources of light overtax the light tolerance of the eye by the damaging effect of ultra-violet rays on the retina. Yellow or amber shooting glasses almost completely eliminate its effect.

    d. Near sightedness is caused by the light rays passing through the lens of the eye and converging before they reach the retina and are starting to radiate.

    e. Farsightedness is caused by the light rays not converging before they reach the retina, therefore are not pinpointed but diffused.

    f. Astigmatism is the passage of light rays through the eye lens, diffusing at random and not converging at any point. Spotty patches of clear and hazy vision is the result.

   3. Correction of defects in vision is necessary. For shooters, even those insignificant defects of vision discovered by tests must be corrected by fitting special glasses for firing. The excessive accommodation necessary for the eye to retain normal acuity is very fatiguing and can lead to relatively early decrease in accuracy.

   4. It is necessary to discuss one more peculiarity of our eyes which has tremendous importance in aiming; the existence of monocular and binocular vision.

    a. Vision with one eye is called monocular and with two eyes, binocular.

    b. Each shooter must determine which is his dominant eye.

NOTE: EXPLAIN THE MANNER IN WHICH THIS IS ACCOMPLISHED. REFER TO CHAPTER II, PAGE 55.

    c. Monocular use of the eyes causes the shooter to close or squint one eye. The dilated pupil of the closed eye affects the shooting eye. The pupil of the open eye seeks to adjust sympathetically and maintain equal acuity in both eyes. This action by the shooting eye reduces its acuity.

    d. With binocular vision, the line of sight is also achieved with one eye. The handicap of the shooting eye endeavoring to adjust to its closed twin is alleviated.

   5. Sharpness of vision is usually determined by the minimum space that we are able to see between two objects. In order for this space to be visibile, it is necessary for at least one retinal element lying between the images of these two points to be stimulated. Thus, the normal sharpness of vision is generally considered to be that at which the eye can distinguish between two visible points at an angle of one minute. This means that the normal eye can distinguish sufficiently clear, for example, a space of .1 inch between the side of the front sight and vertical inside surface of the rear sight notch at a distance of one yard (the approximate distance from the eye to the pistol muzzle and front sight). But the eye of an experienced shooter can distinguish a considerably smaller space between two objects. A number of experiments carried out by specialists attest to the greater accuracy of a shooter's sharpness of vision. For example, the vertical light space between the sides front and rear sight against a white background can be discerned down to the minute width of .01 of an inch at the same distance.

   6. Accurate calling of a shot is dependent upon exact recall of the mental image of the sight alignment at the instant of firing.

   7. The shooter must not use the eye intensely for too long a period when aiming as this tends to tire the eye. The shooter will begin to experience changing degrees of accuracy due to unnoticed errors in sight alignment.

TRANSITION: Not only does the eye help us to maintain correct sight alignment, but plays a great part in the employment of all of the fundamentals. Without a good grip, stance, position to name a few, it would be extremely difficult to have and keep proper relationship of our sights.

 VI. CONCLUSION: (2 min)

  A. Retain Attention: All of us like to gaze upon a beautiful woman or a jewel of great value, or possibly a fine object of art. Why? Because it is very pleasing to the eye and mind.

As shooters, you and I desire also to look at perfect sight alignment each time we extend the weapon toward the target to deliver a shot. Why? Because we feel that we must do our utmost to put that shot or shots in the desired spot on the target. When we do this we know that we have accomplished a feat that will make us the envy of all our fellow competitors. It gives us a good feeling. One that comes only to the master of his thoughts, emotions and body.

B. Summary: Force yourself to align your sights. Keep the front sight centered in the rear sight notch with equal amount of light showing on either side of the front sight. The top of the front sight must be level with the top of the rear sight. You know you can place your point of focus on the front sight. With intense concentration you can keep that focus upon the front sight. By keeping your sights aligned, you will see this.

NOTE: SHOW CHARTS SIX, SEVEN AND EIGHT COMBINED INTO PERFECT SIGHT ALIGNMENT AGAINST A BACKGROUND OF A HAZY BULLSEYE. MOVE THE SIGHT ALIGNMENT COMBINATION WITH A SMALL, ERRATIC MOTION RELATIVE TO THE BULLS EYE. THIS DUPLICATES MINIMUM ARC OF MOVEMENT.

When you see this combination long enough to apply the fundamentals necessary to deliver that shot or shots to the X or ten ring consistently, you will also know that your eyes are performing as well as the mind and the rest of the body.

C. Application: If you apply all you have learned here today you will see an improvement in your scores. You must employ perfect sight alignment in conjunction with the other fundamentals in your practice periods, dry firing sessions and in matches.

D. Closing Statement: You cannot consistently deliver a shot or shots on the target without good sight alignment. With good sight alignment you will post winning scores on the scoreboard. With a string of winning scores on that board, YOU can be the champion. You must generate the desire to excel. You are not wasting time and money by being here. YOU CAN be a winner!

PREPARATION DATA SECTION

I.	TITLE AND NUMBERS:	Fundamentals II: Sight Alignment. (P-2)
II.	TIME ALLOTTED:	Fifty (50) Minutes.
III.	PERIOD OF INSTRUCTION:	One.
IV.	CLASS PRESENTED TO:	Coaches and Instructor-Shooters.
V.	INSTRUCTOR REFERENCE:	The Advanced Pistol Marksmanship Manual.
VI.	INSTRUCTIONAL AIDS:	Blackboard, Chalk, Podium Cards, Podium, Sound Set, Pointer, GTA Charts or 35mm Slides and Projector, Slingshot Device, one 25 Yard Pistol Target (Full Face), one Transparency (Hazy Bullseye), one Transparency (Rear Sight), and one Transparency (Front Sight).
VII.	STUDENT EQUIPMENT & UNIFORM:	Uniform as prescribed.
VIII.	PHYSICAL FACILITIES:	Classroom or outdoor Bleachers.
IX.	PERSONNEL REQUIREMENTS:	1 Principal Instructor, 1 Assistant Instructor, 1 Sound Set Operator.

FUNDAMENTALS III (P-3)

TRIGGER CONTROL

LESSON OUTLINE

I. LESSON OBJECTIVE: To enable coaches and instructor-shooters to gain a knowledge of one of the important fundamentals of marksmanship, trigger control, and to stress the importance of conditioning certain nerve functions of the nervous system by proper training.

II. STUDENT PERFORMANCE OBJECTIVES: As a result of this instruction the student must be able to:

    A. UNDERSTAND the nerve processes involved in trigger control.

    B. UTILIZE factors providing for the correct control of the trigger.

    C. PRACTICE the application of positive trigger pressure.

III. ADVANCE STUDY ASSIGNMENT: The Advanced Pistol Marksmanship Manual, Chapter III, "Trigger Control".

IV. INTRODUCTION: (3 Min)

    A. <u>Gain Attention</u>: In many instances, the hard work devoted to setting up proper stance, position, grip, and perfect sight alignment is ruined at the last second before shot delivery by faulty trigger control. This error in performance is generally caused by concentration being shifted from sight alignment to trigger control shortly before the release of the hammer takes place.

    B. <u>Orient Students</u>:

        1. <u>Lesson Tie-In</u>: In the previous class we discussed the importance of sight alignment. Now we will discuss the remaining factors fundamental to good shooting; trigger control and the importance of positive trigger pressure.

        2. <u>Motivation</u>: To understand how to solve your shooting problems you must have a thorough knowledge of all the fundamentals. You must have a full grasp and working knowledge of trigger control before you can develop into a good pistol shot. A good shooter knows from experience that all the fundamentals must be mastered and coordinated with each other if he is to become a major competitor.

        3. <u>Scope</u>: You will learn to use good trigger control by knowing:

            a. What your nerve processes are and how to condition them for championship shooting by proper training.

            b. How to utilize the factors providing for the correct control of the trigger.

            c. How to apply positive trigger pressure.

V. BODY:

    A. <u>First Student Performance Objective</u>: Students must be able to understand the nerve processes involved in trigger control. (17 Min)

NOTE: SHOW STRIP #1. (NERVE PROCESSES)

    QUESTION: Is it possible to learn to press the trigger causing the shot to fire quickly and exactly when you want it to fire.

    1. There is a definite interval of time between the beginning of the action of the stimulus signal and the beginning of the response movement. (.18 to .25 seconds)

    2. From the beginning of the action to the completion of the response movement it is essential to put into motion certain muscles and restrain the motion of other muscles.

    3. The movement of the trigger finger must be coordinated with correct sight alignment, minimum arc of movement and peak concentration.

    4. Reflexes are the responses to a stimulus; they are divided into two classes.

        a. Unconditioned reflexes are reactions in response to definite external stimuli such as heat, pain or recoil sound and action, etc.

        b. Conditioned reflexes are temporary reactions developed by the influence of your immediate environment.

    5. These are some of the errors made in control of the trigger and means of combating them:

        a. Jerking is the abrupt pressure of the trigger finger accompanied by the sharp straining of all the muscles. Proper dry firing will help to control this.

        b. Holding too long is caused by being excessively cautious and having too slow a reaction on the trigger. This is brought about by fear of a bad shot. Proper dry firing will help this error too.

        c. An incorrectly adjusted trigger can cause much trouble. This should be corrected by a gunsmith. Do not try to alter your trigger control to offset the disturbances caused by a mechanically faulty trigger action.

TRANSITION: You know what your nerve processes are and how to control them. This knowledge must be applied to factors of correct trigger control.

    B. _Second Student Performance Objective_: Students must be able to utilize factors providing for the correct control of the trigger. (12 Min)

    QUESTION: How much initial pressure should be taken up on the trigger?

NOTE: SHOW STRIP #2. (FACTORS PROVIDING FOR THE CORRECT CONTROL OF THE TRIGGER).

    1. Any free movement of the trigger is called slack. This slack must be taken up prior to a light initial pressure of approximately one-fourth of the total required to fire the weapon.

    2. The trigger must be pressed smoothly in such a way as not to disturb any of the established circumstances set up to create conditions for absolute control of the shot.

    3. Your grip must be correct to allow the trigger finger to move freely and properly parallel to the axis of the bore.

4. The trigger finger must be placed so that the trigger can be pressed straight back to the rear.

5. Trigger control must be coordinated with peak visual perception for correct sight alignment, minimum arc of movement and the peak of concentration.

6. The action of pressing positively on the trigger must take place at the time your minimum arc of movement is optimum. You must learn to recognize when your minimum arc of movement is at its' best and approximately how long it will last. From four to ten seconds is average.

TRANSITION: These are the factors providing for the correct control of the trigger. Precise use of these factors is demanded in applying them to positive trigger pressure.

    C. **Third Student Performance Objective:** Students must be able to put into practice the application of positive trigger pressure. (12 Min)

NOTE: SHOW STRIP #3. (APPLICATION OF POSITIVE TRIGGER PRESSURE)

    QUESTION: What is the difference between area shooting and point shooting?

    1. The shooter must endeavor to complete the firing of the shot once the application of trigger pressure has started. This is positive uninterrupted trigger pressure. This will release the hammer without warning, giving you a surprise break. This is basically area shooting If any control factor breaks down during your attempt to fire, bench the weapon and start over.

    2. The interrupted trigger control is used successfully at rare times by only a few of the better shooters. To do this you apply pressure only when the sights are in perfect alignment and your hold is motionless in the center of aiming area. At all other times the pressure is maintained but not increased. This means if you have a bad sight picture, no increase in trigger pressure. Continue to hold the shooting arm extended and wait and wait until it settles again right where you want it. This is point shooting. The usual result is eight tens and two jerked sevens. In interrupted trigger pressure there are things you must watch for:

    a. Thinking about sight picture (target in relation to sight assembly) will break down your concentration on sight alignment. (Relationship of front and rear sight.)

    b. Uncontrolled reflex action may cause you to compensate for the coming jolt of recoil.

    c. Holding too long will occur from breaking concentration and attempting to regain control without bringing pistol down to the bench for a rest and replanning of shot.

    d. Competitive marksmanship is pointed ultimately toward making a shooter proficient under the conditions of combat. Interrupted trigger control would not be an asset under combat conditions. The enemy could walk all over you before you decided to press positive on the trigger.

TRANSITION: Trigger control takes training and practice. Application of positive trigger pressure can be perfected by dry firing.

    VI. CONCLUSION: (6 Min)

    A. **Retain Attention:** Trigger control is very important in the delivery of a good shot. But remember it must be used in coordination with all of the other fundamentals.

B. Summary:

    1. In order for you to attain good trigger control you must be able to:

        a. Understand the nerve processes involved in trigger control.

        b. Utilize factors providing for the correct control of the trigger.

        c. Practice the application of positive trigger pressure.

C. Application: By the proper application of the fundamentals, better scores are inevitable. Remember; there is no secret to winning a match except proper application of fundamentals. If you were to interview outstanding competitors, who have shot 2650 and above about their success in shooting a pistol, these gentlemen would refer you immediately to application of fundamentals.

D. Closing Statement: Before a shooter can become proficient, he must develop, in himself the ability to apply all the fundamentals. The fundamentals cannot be applied fully if the knowledge of them is limited.

PREPARATION DATA SECTION

I.	TITLE AND NUMBERS:	Fundamentals III: Trigger Control (P-3)
II.	TIME ALLOTTED:	Fifty (50) Minutes
III.	PERIOD OF INSTRUCTION:	One
IV.	CLASS PRESENTED TO:	Coaches and Instructor-shooters
V.	INSTRUCTOR REFERENCE:	The Advance Pistol Marksmanship Manual
VI.	INSTRUCTIONAL AIDS:	Blackboard, Chalk, Podium Cards, Podium sound set, pointer, GTA Charts or 35mm slides and projector, Slingshot Device, 25 yard pistol target (full face).
VII.	STUDENT EQUIPMENT & UNIFORM:	Uniform as prescribed.
VIII.	PHYSICAL FACILITIES:	Classroom or outdoor bleachers
IX.	PERSONNEL REQUIREMENTS:	1 Principal Instructor, 1 Assistant Instructor, 1 Sound Set Operator.

TECHNIQUE OF FIRE I

ESTABLISHING A SYSTEM (P-4)

LESSON OUTLINE

I. LESSON OBJECTIVE: To instruct pistol shooters and coaches in the proper method of establishing a system for the delivery of a controlled shot or a string of shots on the target.

II. STUDENT PERFORMANCE OBJECTIVES: As a result of this instruction, the pistol shooter-student should be able to:

   A. ORGANIZE systematic preparation for shooting.

   B. PLAN shot sequence in complete detail.

   C. UNDERSTAND how relaxation delays fatigue and aids muscular control.

   D. DELIVER shot or string of shots on target as planned.

   E. MAKE an analysis of each shot or each string of shots.

   F. MAKE a positive correction (if necessary).

III. ADVANCE ASSIGNMENT: The Advanced Pistol Marksmanship Manual, Chapter IV, "Establishing a System."

IV. INTRODUCTION: (3 Min)

   A. Gain Attention: "For want of a shoe the horse was lost", etc. Pistol shooters lose matches frequently because of overlooking some important factor or item of equipment.

   B. Orient Students:

      1. Lesson Tie-In: Systematic preparation, planning, relaxation, delivery of shot analysis and corrective measures insure that the applied fundamentals will produce good results on the firing line.

      2. Motivation: A systematic operation is essential to insure that the shooter can devote all of his developed skill to shooting his best scores.

      3. Scope: During this period of instruction, a proper approach to the problem of systematic preparation, shot planning, how to relax, delivery of a controlled shot, analysis of a shot and corrective measures, if necessary, will be explained.

V. BODY:

   A. First Student Performance Objective: The pistol shooter must understand how to organize systematic preparation for shooting. (17 Min)

      QUESTION: How much distance in inches does one click move the strike of the bullet on the target at 50 yards with an Eliason sight?

1. Zeroing is accomplished by the shot group method.

    a. Two types of sights are in common use.

       (1) Fixed sight are seldom used in competition.

       (2) Adjustable sights allow quick adjustment of zero.

           (a) Eliason.

           (b) Colt.

           (c) Micro.

           (d) Smith and Wesson.

           (e) Hi-Standard.

           (f) Giles.

           (g) Miscellaneous.

    b. Use the shot group method to properly and quickly zero your pistol

       (1) Zero initially at 25 yds slow fire by the shot group method.

       (2) Confirm zero by timed or rapid fire.

       (3) Adjusting fixed sight.

       (4) "Kentucky Windage."

    c. Mark sights at the 25 yard and 50 yard setting after zeroing.

       (1) Methods of marking.

       (2) Basic sight setting.

       (3) Use of sight adjustment card.

       (4) Use of score book for record of sight setting for all conditions.

    d. Know how to cope with unusual zero changes.

       (1) Check scope.

       (2) Check sights, other external parts.

       (3) Check position and grip.

       (4) Take positive action.

       (5) Check scorebook.

       (6) Check with armorer.

2. Preparation in the Assembly Area.

   a. Physical.

   (1) Personal readiness.

   (2) Limber up before attempting maximum control

   (3) Arrive early at range.

   (4) Check out firing line.

   (5) Check out fit of clothes and shoes.

   (6) Move to area of assigned target.

   b. Mental.

   (1) Stimulate confidence.

   (2) Expect to work hard.

   (3) Prior planning of actions on the firing line.

   (4) Do not be upset by range irregularities.

   (5) Think only of shooting.

   (6) Mentally review shot sequence.

3. Preparation on the Firing Line.

   a. Place shooting box on correct firing point.

   b. Scope your target.

   c. Inspect target for holes

   d. Adjust ear protectors.

   e. Load magazines with proper ammunition.

   f. Check sight blackening.

   g. Locate accessory shooting equipment.

   h. Take a few deep breaths.

   i. Assume a stance.

   j. Check position.

   k. Assume grip on pistol.

   l. Trigger pressure straight to rear.

m. Complete firing line preparation in three minutes.

4. Preparation After the Command "LOAD".

a. Verify stance, position and grip.

b. Load your weapon.

c. Do not engage safety lock.

d. Recheck grip.

e. Recheck for proper target.

f. Relax with pistol at bench rest.

g. Resume mental process of planning shot delivery.

TRANSITION: Planning is theoretically a part of preparation to shoot but the detailed mental review of the successive actions to be performed during the sequence of firing a shot is a necessary prelude to precise execution.

B. <u>Second Student Performance Objective:</u> The pistol shooter should be able to plan shot sequence in complete detail. (6 Min)

QUESTION: What does a planned shot sequence include?

1. Consistently accurate shots are fired by a shot sequence method.

2. The following is the shot sequence used by the USAMTU pistol shooters for slow fire with only minor modifications.

a. Extend arm and breathe.

b. Settle into a minimum arc of movement.

c. Pick up sight alignment in the aiming area.

d. Take up trigger slack - apply initial pressure.

e. Hold breath.

f. Maintain sight alignment and minimum arc of movement.

g. Start positive trigger pressure.

h. Concentrate point focus on front sight.

i. Follow through. (occurs with surprise shot only) (No reflex action)

3. The following is the shot sequence used by the USAMTU pistol shooters for sustained fire with only minor modifications.

a. Extend shooting arm and breathe.

        b.   Find sight alignment.

        c.   Find aiming area on edge of target frame (final deep breath).

        d.   Settle into minimum arc of movement.

        e.   Point focus on front sight (partly release breath).

        f.   Take up slack - apply initial trigger pressure.

        g.   Maintain sight alignment (target faces).

        h.   Start positive trigger pressure.

        i.   Concentrate on sight alignment (first shot is fired).

    4.   Following a system allows the shooter to concentrate on performance.

<u>TRANSITION</u>: When the shooter has completed his preparations and is ready to shoot, there should not be any unnecessary tenseness in his body.

    C.   <u>Third Student Performance Objective</u>: The pistol shooter should understand how relaxation delays fatigue and aids muscular control.   (8 Min)

        <u>QUESTION</u>: How does the shooter relax the proper muscles while shooting?

        1.   Methodically think of relaxing each principal muscular mass of body; neck, shoulders, back, abdomen, buttocks, and upper legs.

        2.   A relaxed muscle does not fatigue and tremble.

        3.   Rest arm after an unsuccessful effort to shoot.

<u>TRANSITION</u>: The relaxed and ready shooter must deliver the shot or string of shots as planned.

    D.   <u>Fourth Student Performance Objective</u>: The pistol shooter should be able to deliver shot or string of shots on target as planned.   (9 Min)

        <u>QUESTION</u>: What are the shooters actions as each fire command is given.

        1.   Example of a system of delivering a sustained fire string with fire commands.

            a.   "ON THE FIRING LINE FOR YOUR FIRST STRING OF RAPID FIRE. WITH FIVE ROUNDS LOAD."

                (1)   You should load at this time and assume your grip as planned.

            b.   "IS THE LINE READY?"

                (1)   Eliminate all thoughts of stance, position, grip. They should be as perfect as you can make them at this period.

c. "THE LINE IS READY."

    (1) Continue your rhythmic breathing and extend your shooting arm.

d. "READY ON THE RIGHT."

    (1) Extend arm, stiff wrist and locked elbow.

    (2) Find sight alignment.

    (3) Breathe deeply and exhale.

e. "READY ON THE LEFT."

    (1) Find aiming area on edge of target frame.

    (2) Final deep breath.

    (3) Settle into minimum arc of movement.

f. "READY ON THE FIRING LINE."

    (1) Partly release breath and hold remainder.

    (2) Point focus on front sight.

    (3) Take up slack - apply initial trigger pressure.

    (4) Maintain sight alignment - all conditions right.

g. Target faces toward shooter - commence firing.

    (1) Start positive trigger pressure.

    (2) Shift concentration to perfecting sight alignment.

h. First shot is fired (Surprise Shot).

    (1) Maintain eye focus (Follow Thru).

    (2) Quick recovery with sights approximately in alignment, and hold approximately in center of aiming area.

    (3) Renew positive trigger pressure.

    (4) Strive to correct errors in sight alignment, but do not delay positive trigger pressure.

TRANSITION: Every shot or string of shots fired has features or faults that should be noticed by the shooter.

    E. <u>Fifth Student Performance Objective</u>: The pistol shooter must be able to make an analysis of shot or string of shots. (9 Min)

QUESTION: If a mistake is made in shooting and no solution is found, what must the shooter do?

1. Analyzing single shots (Slow Fire Worksheet).

    a. Follow through check.

    b. Call shot (Describe sight alignment).

    c. Compare target hit location with shot call.

    d. If shot or call is bad, determine cause.

    e. Watch for error pattern to form.

    f. Did shot break in normal minimum arc of movement?

    g. Did you hold too long?

    h. Did you apply positive trigger pressure?

    i. If you benched weapon on a shot effort, why?

    j. Did you lose concentration? (What did you think about other than sight alignment?)

    k. Did you get a surprise shot break?

    l. Were you worried about results?

Complete and instantaneous shot analysis is a mandatory prerequisite for any improvement in your performance or scores. It is a complete waste of time and ammunition to stand on the line and fire haphazardly without any comprehensive attempt to improve. A mental impression of where each shot went and why, should come at the instant the shot breaks.

2. Analyzing strings of five shots (Rapid Fire Worksheet).

    a. Follow through and proper recovery check.

    b. Shot group call (describe five individual sight alignments).

    c. Compare group location with call.

    d. If shot group or call is bad, determine cause.

    e. Did you get a surprise break on each of five shots?

    f. Was the first shot fired on time?

    g. Was rhythm maintained throughout string including last shot?

    h. Did all shots break in normal arc of movement?

    i. Did you apply positive trigger pressure on each of five shots?

    j. Did you lose concentration during string? (What were you thinking of?)

k. Did you ignore minor errors in hold?

l. Were you worried about results?

    After each five shot string, the shooter should remember each shot as one of five individual sight alignments that enables him to accurately call the shot group.

TRANSITION: After an analysis has been made, positive corrective action must be taken promptly to correct errors in performance. A shooter will never become a champion if he does not analyze and correct mistakes every time a string is fired.

    F. Sixth Student Performance Objective: The pistol shooter should be able to make a positive correction (if necessary). (3 Min)

1. Incorporate into performance on next shot.

2. Prevent recurrence of a poor performance.

3. Shooter must correct errors or lose hope of progress.

VI. CONCLUSION:

    A. Retain Attention: This material showing the correct method of zeroing, shot delivery, analysis, correction and preparation contributes to the improved results the shooter can produce. It is beneficial to the shooter in that it assures him of a system to follow whereby costly mistakes can possibly be eliminated.

    B. Summarize:

    QUESTION: What are the six parts of a system for accurate pistol shooting?

1. Preparation - Physical and Mental.

2. Plan shot sequence.

3. Relaxation.

4. Deliver shot or string of shots on target.

5. Make an analysis of shot or string of shots.

6. Positive correction (if necessary).

    C. Application: These methods of preparation, performance and zeroing will be used in all or in part during the different phases of this course and later when you are shooting under match conditions.

    D. Closing Statement: A shooter who has a well zeroed weapon and has prepared and planned properly for a match has relieved his mind of anxiety during the shooting phase of any of those numerous essentials that may have been overlooked. This will allow him to concentrate on shooting to the full extent of his ability.

PREPARATION DATA SECTION

I.	TITLE AND NUMBER	Establishing a system (P-4)
II.	TIME ALLOTTED:	Fifty (50) Minutes
III.	PERIODS OF INSTRUCTION:	One
IV.	CLASS PRESENTED TO:	Coaches and Instructor Shooters
V.	INSTRUCTOR REFERENCES:	The Advanced Pistol Marksmanship Manual
VI.	INSTRUCTIONAL AIDS:	1 Podium, 1 Sound Set, 4 enlarge standard American targets (25 yds) with shot groups, GTA Charts, or 35mm slides & Projector, 1 worksheet booklet, enlarged Standard American target (50 yds) with sliding shot group. Podium Cards, 1 Blackboard, Chalk
VII.	STUDENT EQUIPMENT AND UNIFORM:	Uniform as prescribed.
VIII.	PHYSICAL FACILITIES:	Outdoor bleachers or classroom for students
IX.	PERSONNEL REQUIREMENTS:	1 Principal Instructor, 1 Assistant Instructor, 1 Sound Set Operator.

TECHNIQUE OF FIRE II

SLOW FIRE (P-6)

LESSON OUTLINE

I. LESSON OBJECTIVE: To give the shooter a technique for control of slow fire by bringing about a better understanding of the employment of the fundamentals. The secret of winning pistol matches, if any mysterious formula exists at all, is the knowledge of properly controlled employment of all the fundamentals in the firing of one accurate shot.

II. STUDENT PERFORMANCE OBJECTIVES: As a result of this instruction the pistol shooter-student should be able to:

    A. CONTROL and coordinate the employment of the fundamentals in firing each shot of slow fire.

    B. DEVELOP a technique for slow fire control.

    C. RECOGNIZE and correct common deficiencies in control.

    D. COMPENSATE for adverse shooting conditions.

    E. USE comprehensive training methods.

III. ADVANCE ASSIGNMENT: The Advanced Pistol Marksmanship Manual, Chapter VI, "Technique of Slow Fire" and Chapter V, "Coordination of Control Factors".

IV. INTRODUCTION: (3 Min)

    A. <u>Gain Attention</u>: A technique can be defined as a highly specialized method of performing a specific, complex operation. A technique for control of slow fire without a doubt requires specialization and is based on a complex system of organization and coordination for accurate delivery of each shot.

    B. <u>Orient Students</u>:

        1. <u>Lesson Tie-In</u>: Up to this point you have had instruction in stance, position, grip, breath control, trigger control, and sight alignment. It is necessary to learn to employ all these fundamentals in coordination and develop a technique for slow fire.

        2. <u>Motivation</u>: Generally, in any pistol match more points are dropped at the 50 yard line slow fire than in timed or rapid fire. Here is an example of how a fifty percent improvement in either stage of fire will affect your total score in the National Match Course.

SLOW	TIMED	RAPID	TOTAL
84	98	94	276
Improve 25 yard scores by 50%			
84	99	97	280
Improve slow fire score by 50%			
92	98	94	284

3. _Scope_: The shooter can improve his slow fire performance by:

    a. Controlling and coordinating the employment of the fundamentals in firing of each shot of slow fire.

    b. Developing a technique of slow fire control.

    c. Recognition and correction of common deficiencies in control.

    d. Learning to compensate for adverse shooting conditions.

    e. Using comprehensive training methods.

V. BODY:

    A. _First Student Performance Objective_: The slow fire shooter must understand how coordination of the control factors will bring about control of employment of the fundamentals in firing each shot of slow fire.   (12 Min)

    QUESTION: What are the control factors?

NOTE: POINT OUT EACH OF THE FOLLOWING FACTORS ON CONTROL FACTOR TRAINING AID.

    1. The fundamentals of stance, position, grip and breath control, when employed properly, will give the shooter the ability to hold the pistol almost motionless within the center of the aiming area on the target. This is establishing the minimum arc of movement.

    a. Governs the basic size of shot group if sights are kept uniformly in perfect alignment.

    b. The duration of optimum hold or minimum arc of movement is approximately three (3) to six (6) seconds.

    2. Sight alignment is the relationship of the front sight to the rear sight while aiming.

    a. Any misalignment of the front and rear sights will cause an angular shift error of three (3) inches on the target at fity (50) yards for each one hundredth (.01) inch of deflection.

    b. The point of focus is the front sight. The duration of optimum visual perception - the ability of the human eye to maintain the point focus, is limited to a period of six (6) to eight (8) seconds.

    3. Positive trigger control is the act of committing the pressure on the trigger to completion of the firing of the shot after application is once started.

    a. Trigger slack and initial pressure is taken up before positive trigger pressure is started.

    b. Uninterrupted positive trigger pressure is not applied until the other control factors are settled and are as near perfect as the shooter can set them up.

       c. The duration of positive trigger pressure in firing the pistol varies from two (2) to five (5) seconds because of the care needed to maintain smoothness of application of constantly increasing pressure straight to the rear and not disturb sight alignment and observe the limitations or time in which control factors are optimum. Smooth trigger pressure has been applied in as short a period as one second in rapid fire.

       d. The hammer should fall and fire the shot as a surprise. Any anticipation of the release of the hammer will set up reflex muscular action in the shooting arm and hand and spoil the accuracy of the shot by disturbing sight alignment.

    4. Intense mental concentration of the degree necessary to think only of maintaining the front and rear sights in exact alignment will function for the average shooter at an optimum level for approximately three (3) to six (6) seconds.

    5. To exercise maximum control of shooting a pistol, all of the control factors must be coordinated to be at or very near their optimum state simultaneously. Coincidence should come during the three (3) to six (6) second interval when mental concentration is at peak intensity.

<u>TRANSITION</u>: When all of the control factors are in coordination, employment of the fundamentals is under a high degree of control. A technique must be developed that will maintain consistent control over coordination of employment of the fundamentals once it is attained. A method must be found which will assure constant renewal of perfect employment of each of the fundamentals.

  B. <u>Second Student Performance Objective</u>: The shooter must develop a technique for slow fire control. (10 Min)

    <u>QUESTION</u>: When all conditions are right, how long should it take a shooter to fire a shot of slow fire?

    1. A shooter should form a habit of firing a few "Dry" shots before beginning to fire. This action brings into play, if executed properly, all of the control factors that will be utilized in match firing. If the coordination is inconstant or the employment of one or more of the fundamentals is faulty, immediate corrective steps can be taken to smooth out the operation. Mastery of the employment of the fundamentals is a result of constant practice and extensive match firing experience after the correct method is learned.

    2. Great care is one of the mainstays of the control of slow fire shooting. Many of the poorly controlled shots are the result of various degrees of carelessness. A habit or an infrequent compromise in the form of accepting the conditions the shooter sets up for a shot as being almost good enough is a form of carelessness. An important sector of slow fire control lies in accepting no compromise on perfection. Faultless coordination is a result of countless repetitions of a perfect performance.

    3. Patience is of extreme importance to sustaining an effective technique for control of slow fire. Without patience the shooter may disrupt an otherwise perfect performance an instant from successful completion.

    4. Oversighting or holding too long is a consequence of over-caution brought on by the shooter's excessive fear of getting a bad shot. Slow pressure on the trigger arises when coordination is difficult. This difficulty is manifest when the process of inhibition predominates over the process of stimulation. Usually the shooter simply will not press positively on the trigger even though the other control factors are optimum. Lack of confidence in his ability to

maintain the control factors static long enough to fire an accurate shot is the underlying cause of this reluctance.

        5. The most efficient manner in which the shooter can impart every ounce of his skill to his shooting is by careful organization of his approach to exerting maximum control. For this a system is needed. When organized, this system of delivering an accurate shot on the target should be used uniformly for each shot. The detailed method of developing a personal system is outlined Chapter IV, "Establishing a System", Volume I.

        6. In determining pace or tempo of shooting, experience has shown that the most effective approach is to shoot rather rapidly or within a maximum of six (6) seconds after settling. This brief time limit is necessary if the technique of shooting follows the premise of coor dina ion of control factors in employment of the fundamentals, covered in the first student performance objective, this chapter.

        7. The time spent between each shot, preparing and planning is limited only by the time allowed for the ten shot string of slow fire. The shooter's energy must be expended with care by taking sufficient rest time between each shot and each ten shot string by observing a definite pacing and rhythm of operation. If the weather or other conditions are not favorable, the shooter may have to alter his tempo and approach to control to meet prevailing conditions.

<u>TRANSITION</u>: Slow fire control produces successful results only when flaws in performance are at a minimum. The shooter should be familiar with the more common deficiencies in control and also how to avoid and correct them.

    C. <u>Third Student Performance Objective</u>: The pistol shooter-student should know the common points of deficiencies in slow fire control.   (10 Min)

    <u>QUESTION</u>: What is the difference between a jerk and a heeling of the shot?

        1. & 2. <u>Jerk or Heel</u>: The abrupt application of pressure either with the trigger finger alone or in the case of heeling, pushing with the heel of the hand at the same time. Apply pressure to the trigger straight to the rear and wait for the shot to break. Anticipation can cause muscular reflexes of an instant nature that so closely coincide with recoil that extreme difficulty is experienced in making an accurate call. Anticipation is the same as flinching.

        3. <u>Vacillation</u>: This is a mental fault which results in your acceptance of various minor imperfections in your performance which you could correct if you worked out a plan of action. The end result is usually that you hope to get a good shot. The method of correcting this fault is to develop a comprehensive plan and follow it without deviation until you can make improvements in it.

        4. <u>Anxiety</u>: You work and work on a shot, meanwhile building up in your mind doubt about the possibility of the shot being good. Finally you shoot just to get rid of that particular round so you may work on others. Napoleon arrived at Waterloo traveling the same bumpy road.

        5. <u>Not Looking at the Sights</u>: This quite frequently is listed as "looking at the target." A shooter may be focusing his eye on neither the sights nor the target, but since he does not see the target in clear focus he assumes he is looking at the sights. You must concentrate on sight alignment.

        6. <u>Loss of Concentration</u>: If the shooter fails in his determination to apply positive pressure on the trigger while concentrating on the front sight, his prior determination to apply the positive pressure needs to be increased and reemphasized.

7. <u>Holding too Long</u>: Any adverse conditions that disturb a shooter's ability to "hold" will cause him to delay his positive application of trigger pressure waiting for conditions to better. The disturbing factor about this is that you will do it unconsciously; therefore, you must continuously ask yourself, "Am I trying to freeze all arm movement momentarily so I can get a shot off quickly before movement is resumed?"

8. <u>Overcorrection</u>: Maintaining control of your shooting is a continuous battle. The battle builds tension. Tension tightens the muscles and finally the abrupt motions made in compensation for errors cause the shooter to go beyond the desired area and deliver shots in exactly the opposite place from where the error was causing him to shoot originally. Smoothly coordinated actions are best assured by the relaxed, confident and carefully planned approach.

9. <u>Lack of Follow Through</u>: Follow through is the subconscious attempt to keep everything just as it was at the time the shot broke. In other words you are continuing to maintain concentration on sight alignment even after the shot is on the way. Follow through is not to be confused with recovery. Merely recovering and holding on the target after the shot is fired is not indication that you are following through.

10. <u>Match Pressure</u>: (See Chapter VIII, "Mental Discipline") If there are 200 competitors in a match, rest assured that there are 200 shooters suffering from match pressure. What makes you think you are so different? You should exert all your mental energy toward planning and executing the fundamentals correctly. Your shooting match pressure will become controllable and your competitors will congratulate you on your fine performance. There are obviously a multitude of causes for bad shots. We have listed those most frequently found. It is not intended to be a complete list nor is it intended to provide the shooter with a convenient list of bad habits. It is, however, intended to assist the shooter in finding the source of his trouble.

<u>TRANSITION</u>: Ability to correct deficiencies in performance under ideal conditions points the shooter toward the more difficult conditions under which shooting must be controlled.

D. <u>Fourth Student Performance Objective</u>: The pistol shooter-student should know how to cope with the wind and other adverse shooting conditions. (5 Min)

<u>QUESTION</u>: If the wind is blowing hard and the shooter cannot hold in the bullseye area, should he shoot anyway?

1. Wind shooting is conducive to jerking the trigger. This is true because as the arc of movement increases, the shooter develops a tendency to relax his trigger pressure. He is waiting for a more stable sight picture. His concentration on sight alignment will diminish and he will make an effort to set the shot off on the move as the sights pass the vicinity of the target center. The obvious answer is to, first wait for a lull in the wind; next, concentrate as one normally does on sight alignment and as a minimum arc of movement is achieved, start a constantly increasing positive pressure on the trigger until the shot is fired. Do not continue the hold during extreme gusts. Always take advantage of a chance to rest. Each attempt to fire a shot should be made with a firm resolve to align the sights and to apply constantly increasing trigger pressure until the shot is fired. The surprise shot continues to be the indicator, even under these conditions, of whether you are applying the fundamentals. Your shot group will be larger as a result of the increased arc of movement but the wild shots resulting from faulty sight alignment, flinching, jerking and over-correction will be minimized. Extensive practice under wind conditions is not recommended but enough firing should be conducted under those conditions to prevent a stampede to the nearest wind shelter when a wisp of air movement stirs the pine tops.

2. Adverse weather conditions such as cold, hot or rainy weather or extreme light conditions pose problems that can be solved in much the manner as in wind shooting. Be determined to adhere to the fundamentals and ignore as much as possible the distractions that are demoralizing to the competition.

3. Light condition varies from extremely bright to very dim and the shooter must keep a record of the light conditions on every range fired on in his score book. Some competitors are affected more by changes in light than others. A note should be made as to how much his zero changes in the different light conditions. Sights should be blackened with care on bright days. As a part of shooting accessories you should have both amber and green shooting glasses not only for light conditions but for protection against against oil, wind and empty brass. Firing from an uncovered firing line usually requires different sight sittings than the firing from under a shed. Ammunition should be kept out of the sun as its accuracy is affected if it is exposed to the direct rays of the sun.

4. The major portion of our accomplishments on the firing line stems from our mental capacity to face up to the out-of-the-ordinary and parlay these conditions into a winning margin. Poor conditions must never become an excuse to quit or compromise and consequently deliver a poor performance. Good scores are produced by hard work in the application of the fundamentals regardless of the conditions. Proper application of the fundamentals remains the most important factor in shooting winning scores under adverse conditions.

TRANSITION: The proper use of the control factors in employment of the fundamentals necessitates the following of a comprehensive method of training to analyze and perfect the technique chosen by the shooter.

E. <u>Fifth Student Performance Objective</u>: The pistol shooter-student should be familiar with the various training methods that will improve his overall performance potential.

(7 Min)

QUESTION: If match competition is available each week, should the shooter participate?

1. Frequent shoulder-to-shoulder competition and regularly scheduled practice on the range with shooters who approach the problem of improving their shooting with enthusiasm and a serious, determined attitude is the most effective method of accelerating your development as a top competitive shooter.

2. To be most effective, each practice session must have a goal. You should approach the training period with the idea that you are going to distinctly improve one specific aspect of your shooting technique and at the same time continue the general improvement evidenced by your ability to employ the fundamentals more effectively as your development as a coming champion progresses.

3. To improve your ability to deliver each slow fire shot quickly and accurately, we advise a practice session of about ten rounds delivered in the following manner. Adjust the target turning mechanism to face the target and turn it away after approximately three (3) seconds. Use your normal preliminary preparation with maximum attention on delivering the first shot without hesitation as the target turns. Fire one shot only. Repeat the exercise ten times with sufficient time between shots to allow for analysis and mental preparation.

4. To improve your ability to maintain a point focus on the front sight place a target face on the frame backwards so that no bull's-eye or aiming point is visible. Assume your stance, position and grip with meticulous attention to detail. Without a <u>point</u> to aim at you will find that you must trust your stance and position to maintain an acceptable arc of movement.

You will find it easier to apply the fundamentals and find that you can deliver the shot with amazing accuracy. Sight alignment can be maintained with a startling degree of control and assurance. This is because the distracting effects of having an exact point to aim have been eliminated. You have no way of knowing when a perfect hold occurs. You simply accept minor errors in hold caused by your minimum arc of movement and go ahead and follow your plan of delivery of the shot.

       5. One word of advice: Avoid training and shooting alone. Use a training program that duplicates as near as possible the competitive atmosphere of a match. Develop and use a comprehensive plan that gives you the ability to employ the fundamentals most reliably under pressure and continually strive for improvement. The shooters' Slow Fire Work Sheet will provide this guidance.

       6. Dry firing practice should be conducted with the same careful attention to detail as live ammunition practice.

VI. CONCLUSION: (3 Min)

    A. <u>Retain Attention</u>: The great slow fire shooters do not hope, pray and wish for Providence to smile on their efforts in pistol competition so they can turn in a favorable result without knowing exactly how it happened. Every thought and action is planned deliberately and executed with care and precision. Their technique of slow fire technique is a tangible, workable method that enables them to enter competition with confidence.

    B. <u>Summary</u>:

       1. The shooter should be able to control and coordinate the employment of the fundamentals in firing each shot of slow fire.

       2. The shooter must develop a personal technique of slow fire control.

       3. The shooter must recognize and correct the common deficiencies in control.

       4. The shooter must learn to compensate for adverse shooting conditions.

       5. The shooter must use a comprehensive method of training.

    C. <u>Application</u>: A coach or a shooter must develop a coordinated technique for the systematic application of the fundamentals to obtain winning scores.

    D. <u>Closing Statement</u>: There are no gimmicks or special pills that boost your slow fire scores. The answer is a lot of hard work, care, patience, coordination, analysis of errors, and positive correction to prevent any past errors from occurring again and lowering your performance level.

<u>PREPARATION DATA SECTION</u>

I.	TITLE AND NUMBER:	Technique of Slow Fire (P-6)
II.	TIME ALLOTTED:	Fifty minutes
III.	PERIOD OF INSTRUCTION:	One (1)
IV.	CLASS PRESENTED TO:	Coaches and Instructor-Shooters
V.	INSTRUCTOR REFERENCES:	Advance Pistol Marksmanship Manual

VI.	INSTRUCTIONAL AIDS:	1 Podium, podium cards, 1 sound set, blackboard, chalk, eraser, GTA charts or 35mm slides and projector, Slowfire worksheet, mechanical training aid as depicted in diagram "Coordination of Control Factor, Chapter V, Volume I
VII.	STUDENT EQUIPMENT AND UNIFORM:	Uniform as prescribed
VIII.	PHYSICAL FACILITIES:	Classroom or bleachers
IX.	PERSONNEL REQUIREMENTS:	1 principal instructor, demonstrators and assistant instructors as needed, 1 sound and projector operator.

TECHNIQUE OF FIRE III

SUSTAINED FIRE (P-7)

LESSON OUTLINE

I. LESSON OBJECTIVE: To aquaint the pistol coach-instructor-shooter with an understanding of the techniques of control of employment of the fundamentals in the sustained fire stages and the use of training methods to overcome deficiencies in control and the effect of adverse conditions.

II. STUDENT PERFORMANCE OBJECTIVES: As a result of this instruction the pistol shooter-student should be able to:

A. UNDERSTAND the necessity for control and coordination of employment of the funda mentals of pistol shooting in the sustained fire stages.

B. MAKE use of all known factors having a bearing on developing a technique of sustained fire control.

C. ANALYZE the causes of common deficiencies in control of sustained fire.

D. EXPLAIN how the application of proper training methods will improve performance in sustained fire.

E. EXPLAIN how to overcome the effects of wind and other adverse conditions in the sustained fire stages.

III. ADVANCE ASSIGNMENT: The Advanced Pistol Marksmanship Manual, Chapter VII, "Sustained Fire".

IV. INTRODUCTION:

A. Gain Attention: The Pistol shooter must approach the rapid fire stage with the confidence and assurance that derives from having fired an unlimited number of successful strings of sustained fire.

B. Orient Students:

1. Lesson Tie-In: We have previously discussed the coordinated employment of the fundamentals and how they apply to slow fire. The same fundamentals apply in timed and rapid fire plus the factors of recovery and rhythm.

2. Motivation: Can you consistently shoot good timed and rapid fire scores? Have you ever had a chance to win a match and then blown up on Rapid Fire? Timed and rapid fire can be stumbling blocks, especially if attempted in a haphazard manner. However, through the development of proper techniques and careful planning, you can improve and become more consistent in your performance.

3. Scope: The shooter must be familiar with all the factors that contribute to his control of sustained fire stages.

a. Coordinated control of the employment of the fundamentals.

       b. Develop a technique of sustained fire control.

       c. Analyze the causes of common deficiencies in sustained fire control.

       d. Apply proper training methods.

       e. Overcome effects of adverse conditions.

V. BODY:

    A. <u>First Student Performance Objective</u>: The Pistol Shooter-Student should understand coordinated control of Employment of the Fundamentals. (5 Min)

<u>QUESTION</u>: What are the fundamentals of pistol marksmanship?

       1. Same fundamentals as in slow fire technique.

          a. Minimum arc of movement.

          b. Sight alignment.

          c. Trigger control.

       2. Recovery should be natural and uniform.

       3. Rhythm is the result of proper execution of a planned sequence of action.

<u>TRANSITION</u>: When an understanding of how coordinated control of employment of the fundamentals is achieved, a technique of maintaining control of sustained fire must be developed.

    B. <u>Second Student Performance Objective</u>: The Pistol Shooter-Student must develop a Technique of Sustained Fire Control based on thorough preparation, planning string, relaxation, controlled delivery of string and analysis and correction of errors made. (12 Min)

       1. Find aiming area on edge of target frame in line with position of bullseye when target is faced.

       2. Stiffen your shooting arm for solid arm control.

       3. Establish eye focus on front sight.

       4. Maintain eye focus on front sight throughout string.

       5. Apply positive pressure on trigger as targets turn.

       6. Shift concentration to sight alignment the instant positive trigger pressure has started.

       7. Maintain head position.

       8. Reestablish sight alignment during recovery without focus shift.

       9. Recovery must be natural, uniform and quick.

QUESTION: Why must recovery be natural and uniform?

    10. Reestablish positive trigger pressure during recovery from recoil.

    11. Attempt to correct errors in sight alignment after recovery but do not delay prompt, rhythmic reapplication of positive trigger pressure to fully accomplish this.

    12. Each shot should have a major effort expended to attain perfect sight alignment.

    13. Good rhythm indicates coordinated control of the employment of fundamentals.

    14. Many factors in technique of slow fire are applicable to sustained fire control.

TRANSITION: Before and during the firing of a sustained fire string, errors in performance may be committed. The shooter must familiarize himself with the more common of these errors and learn to analyze and correct all of those committed if progress is to be maintained.

    C. Third Student Performance Objective: The Pistol Shooter should be able to analyze common deficiencies in control of sustained fire. (10 Min)

QUESTION: How does a faulty recovery affect performance?

    1. Understand how lack of follow through affects control.

    2. Slow and faulty recovery means unnecessary delay and requires correction before each shot.

    3. Calling the shot group accurately requires a visual memory of five sight alignments.

    4. Understand how lack of rhythm indicates failure of coordination in control of employment of fundamentals.

    5. Do not try to correct minor errors in hold.

    6. Understand how lack of a system makes it difficult to repeat a good performance

    7. Incomplete shot group analysis allows errors to remain in performance.

    8. Lack of positive correction causes error to continue damaging performance.

    9. Analyze why you are shooting well.

    10. Overeating during the shooting day will penalize the ability to shoot with maximum control.

    11. Inability to control mental processes indicates a failure to maintain motivation to apply plan of action.

    12. Concentration on sight alignment breaks as target faces indicates a lack of continuity in developing and applying plan of action.

TRANSITION: The perfection of sustained fire technique and elimination of errors in performance is dependent upon a training program that provides sound advice, allows time for free practice and assistance in analysis and correction of errors.

    D. Fourth Student Performance Objective: The Pistol Shooter-Student must learn and apply training methods that will improve performance in sustained fire.   (10 Min)

QUESTION: In what ways will dry firing improve performance?

    1. Competition will accelerate your development as a top competitive shooter.

    2. The shooter must have a specific objective for each practice session.

    3. The use of first and second shot drill to enable prompt starting of string as targets face.

    4. Rhythm and sight alignment exercises will improve ability to maintain sight alignment and timing.

    5. Avoid training and shooting alone. A coach or even another shooter can be of great help.

    6. Dry firing practice will improve the shooters coordination, uniform control of employment of the fundamentals, eye focus, analysis and correction of errors, etc.

TRANSITION: Mastery of a technique of controlled employment of the fundamentals under ideal conditions will of necessity be modified when the wind starts to blow.

    E. Fifth Student Performance Objective: The Pistol Shooter must know how to overcome effects of wind and adverse conditions during the sustained fire stages.   (5 Min)

QUESTION: Is it wise to practice frequently in the wind?

    1. Wind shooting is conducive to jerking the trigger and unless the following is observed, control is erratic at best.

        a. Attain a minimum arc of movement.

        b. Maintain sight alignment.

        c. Apply positive trigger pressure.

        d. Achieve rhythm.

        e. Extensive practice is not advisable under windy conditions.

    2. Compensate for other adverse weather conditions.

        a. Rain. Carry a raincoat to every match.

        b. Cold. Have warm clothing available.

        c. Hot. Wear loose, lightweight comfortable clothing.

        d. Effect of temperature on shot dispersion.

3. Compensate for changing light conditions.

    a. Dim.

    b. Bright.

4. Mental Attitude must be one of accepting the existing conditions and work to compensate as much as possible for the enlarged arc of movement.

VI. CONCLUSION:

A. <u>Retain Attention</u>: To shoot good timed and rapid fire scores, you must first know all the fundamentals of your shooting control. You must put all of these factors into coordinated action in practice. After a string is fired, analyze your performance so you can apply the necessary corrective action. Your performance will need periodic correction to assure constant improvement.

B. <u>Summary</u>:

1. Understanding coordinated control of employment of the fundamentals.

2. Develop a technique of sustained fire control.

3. Analyze the causes of common deficiencies in control.

4. Knowing how the application of proper training methods will improve performance in sustained fire.

5. Knowing how to overcome the effects of wind and adverse conditions during the sustained fire stages.

C. <u>Application</u>: Whether you are coaching a shooter or doing the shooting yourself, you will find a planned sequence of action is the technique necessary for precise control. Organization results in being able to successfully repeat a good peformance.

D. Closing Statement: The many things that contribute to precise control during the firing of the five shot string of timed or rapid fire do not just happen by chance but are carefully planned and executed by the shooter. Successfull repetition of a controlled five shot string is possible only by having complete knowledge of how it is done properly.

<u>PREPARATION DATA SECTION</u>

I.	TITLE AND NUMBER:	Technique of Sustained Fire (P-7)
II.	TIME ALLOTTED:	50 Minutes
III.	PERIOD OF INSTRUCTION:	One
IV.	CLASS PRESENTED TO:	Coaches and Instructor Shooters
V.	INSTRUCTOR REFERENCES:	The Advanced Pistol Marksmanship Manual
VI.	INSTRUCTIONAL AIDS:	1 Podium, 1 Pointer, 1 Blackboard, Chalk, Podium Cards, GTA Charts or 35mm Slides and Projectors, Rapid Fire Worksheet (Handout)

VII. STUDENT EQUIPMENT AND
UNIFORM: Uniform as prescribed

VIII. PHYSICAL FACILITIES: Outdoor bleachers or classroom.

IX. PERSONNEL REQUIREMENTS: 1 Instructor, 1 Assistant Instructor, 1 Sound set operator.

MENTAL DISCIPLINE (P-8)

LESSON OUTLINE

I. LESSON OBJECTIVE: To acquaint the competitive shooter with the fact that precise, uniform control of the physical skills employed in pistol shooting is dependent upon the degree of mental control with which the shooter is able to govern his action on the firing line. POSITIVE THINKING and CONFIDENCE are the key words.

II. STUDENT PERFORMANCE OBJECTIVES: As a result of this instruction the pistol shooter-student must be able to:

   A. UNDERSTAND why mental discipline is essential to advance pistol marksmanship.

   B. DEVELOP mental discipline, confidence and the ability to think positively.

   C. EXPLAIN why negative thinking prevents a shooter from winning.

   D. UNDERSTAND the causes, effects and control of match pressure.

   E. REDUCE the effects of tension and attain relaxation under conditions of stress.

III. ADVANCE ASSIGNMENT: The Advanced Pistol Marksmanship Manual, Chapter VIII, "Mental Discipline".

IV. INTRODUCTION: (3 Min)

   A. <u>Gain Attention</u>:

       1. The members of this marksmanship class represent a typical cross-section of the personnel in the armed forces and the general adult population. There is a wide variety of mental capacities from infantile to near genius. There is only one genius present, your charming instructor.

   B. <u>Orient Students</u>:

       1. <u>Lesson Tie-In</u>: In the preceding instruction the shooter has received information on pistol shooting fundamentals, technique of employment of the fundamentals by coordination of the control factors. This instruction will cover the development and utilization of a mental discipline that will permit the shooter to maintain his control of employing the fundamentals.

       2. <u>Motivation</u>: Properly utilized mental discipline makes it possible for the shooter to maintain control of employment of the fundamentals. This ability is reflected in the shooter by confidence, positive thinking and the proven ability to repeat the delivery of a successful shot. This is the sure way to achieving a more perfect performance and victory in competition.

       3. <u>Scope</u>: You attain mental discipline by having knowledge of:

           a. Why mental discipline is essential to advanced pistol marksmanship.

           b. How to develop mental discipline, confidence and the ability to think positively.

        c. Why negative thinking prevents a shooter from winning.

        d. The causes, effects and control of match pressure.

        e. How to reduce the effects of tension and attain relaxation under conditions of stress.

NOTE: ASK IF THERE ARE ANY QUESTIONS CONCERNING THE ADVANCE ASSIGNMENT.

   V. BODY:

     A. <u>First Student Performance Objective</u>: The pistol shooter-student must understand why mental discipline is essential to advanced pistol marksmanship. (5 Min)

        QUESTION: Why is mental discipline essential to advanced pistol marksmanship?

        1. Mental control has become essential to advanced marksmanship because mastery of the physical skills alone does not provide the uniform, precise performance necessary to compete at the highest level. Too little emphasis is placed on teaching how and what to think.

        2. Mental discipline provides the grasp the shooter has to have to maintain his confidence, positive thinking, and ability to repeat a successful performance. It further provides continued interest which is stimulated by the desire to improve and also the ability to channel and sustain mental effort. It will help thought and action and avoid over-confidence, pessimism and exposure to conditions that will disrupt his mental tranquility.

        3. Mental discipline provides the emotional stability so necessary to the development of the champion shooter. Confidence in his ability and mastery of the basic skills combine to produce a dependable performance under all degrees of stress.

        4. The self-control attained by the advanced pistol shooter pays off not only in better match scores, but also in combat situations by the calmness and resolution in using his weapon to kill with nearly every round fired.

TRANSITION: The essential nature of mental discipline to advance pistol marksmanship requires that the shooter develop this factor as part of his shooting skill.

     B. <u>Second Student Performance Objective</u>: The Pistol shooter-student must be able to develop mental discipline, confidence and the ability to think positively. (12 Min)

        QUESTION: How is mental discipline developed?

        The continuously repeated, successful execution of a step-by-step, completely planned approach to the firing of each shot as pertains to the physical acts and certain elements of mental control involved, results in the gradual development of a mental discipline. The proper degree of mental discipline restricts the thoughts and actions during shooting to an established pattern from which there will be few deviations. Adopt the positive attitude and make up your mind how you are going to fire the shot. Psychologists have determined that there are four basic methods of responding to a problem. Two methods are positive and classified as either direct or indirect. Two methods are negative and classified as either retreat or evasion.

        1. Positive response to a problem.

           a. The direct, positive approach. This is the self-confident, self-sufficient, direct, positive attack that realistically faces the facts, analyzes them, identifies the obstacles to a successful solution and proceeds to grapple tenaciously with them until the solution is found. You know what you want to accomplish and you take direct steps to attain it.

        b. The indirect, substitute or compromise approach. Small, diffident, tentative, indirect actions in which sidestepping leads to seeking short-cuts and when the probable solution is tried, there is much feverent hoping that the fates are on your side. You are only hinting and probing instead of stating definitely what you need to do.

    2. Negative Response to a Problem.

        a. The negative retreat. The failure to give the honest try to see what you are capable of accomplishing. Surrendering without an honest attempt. The flight habit can become chronic. This is the man that cannot accept the responsibility for a mistake or failure. A bad shot produces excuses.

        b. Evading the issues. Evasion is the lack of incentive. Why, is the approach, Why do I have to do better than anybody else? If the desire to excel is not there, you will never aimlessly or otherwise achieve the degree of accomplishment that crowns the champions.

    3. Analyze the Problem.

        a. Psychologists have discovered that one of the chief reasons for difficulty in the solution of a problem is inability to soundly analyze. Pose a clearcut plan of action in full array, facing the specific difficulties and where faced with a particular difficulty, make a determined effort to break it down. If it is identified, there is a solution for it because there are shooters on your team or some other team that are operating without this specific problem putting a brake on their performance. Air it out. A communal pondering session will break it wide open.

        b. There is a four-point system of analyzing and solving specific problems. It reduces the whole big problem to many specific small ones. Head four columns on a sheet of paper with the following titles: one, '<u>STEPS IN THE PLANNING</u>' of control for firing an accurate shot; two, '<u>SPECIFIC DIFFICULTIES</u>' in performing each step of the shot plan; three, '<u>SUCCESSFUL SOLUTIONS</u>' to each of the steps that are being performed satisfactorily; four, '<u>DEGREES OF SUCCESS</u>' with those difficult steps that aren't working out too well. Refer here to Chapter IV, "Establish a System" and follow the complete sequence of firing a shot.

        c. The positive action approach requires that we be specific, that we have a definite plan, that we support the plan by consistent use of each step of the plan, that we be persistent in the face of difficulty in execution of any step of the plan, not rest until a solution is reached and finally that we be on guard against compromise and negative thoughts. The positive approach to overcoming obstacles in our ability to exert mental control over our actions can become automatic. The power of positive thinking leads to confidence.

    4. Confidence results from repeatedly bringing under control all the factors that create conditions for an accurate shot. An accurate shot is one that hits the target within the shooter's ability to hold. People have been telling you for years that you must have confidence to shoot well. Confidence in what? How do you get it? How do we keep it once we put our hands on it?

        a. First and foremost you must have confidence in the fundamentals of advanced pistol marksmanship that you use. You must be convinced that if you control their employment correctly, you will achieve excellent results.

        Nothing can be more undermining than to attempt any task with wishy-washy ideas about how to accomplish it. You must believe and preferably prove to yourself, for example, that sight alignment is vastly more important than sight picture. Believing in correct rhythm and your ability to execute the same, is the greatest deterrent to anxiety in rapid fire. The techniques of employment of the fundamentals that you have proven sound and dependable by experience are not going to change suddenly to unreliable factors because of match pressure.

        b. Confidence in yourself and your ability to execute these proven fundamentals correctly. You have proven your degree of ability to do this in your practice sessions. Go ahead and do it in the big match. To the timid and hesitating, everything is impossible, because it seems so.

        c. Think big! Think positive! "I will do it," and you will succeed. However, as soon as you admit the slightest possibility of failure, so long as there is an influence in your mind that is preventing you from putting all your energies into your task, your success is questionable.

        d. It has been said innumerable times that a pistol shooter must have an open mind, implying that we must have the ability to accept new ideas. What we should also strive for is a mind that is open to positive thoughts and completely closed to those of negative vein.

    5. Channeled mental effort resists the tendency of the mind to drift during the period when intense concentration on sight alignment is essential.

        a. Channel Mental Effort relentlessly toward the final act, as does the trigger pressure that releases the firing mechanism without disturbing sight alignment.

        b. Complete Exclusion of Extraneous Thoughts for a brief period (three to six seconds) is necessary for controlled delivery of the shot.

        c. Prior Planning of the Sequence of Action is necessary to deliver a controlled shot on the target and gradually enables the shooter to sustain concentration for a longer period

        d. Careful Planning of a Sequence of Events closes the mind to other thoughts. Example: If a prior plan is made to apply positive trigger pressure when sights are in alignment and the arc of movement is at the minimum; uninterrupted, positive trigger pressure becomes almost involuntary.

        e. Coordination of Thought and Action is the result of experience obtained through extensive practice and match shooting where the same satisfactory plan of action is followed repeatedly.

TRANSITION: Knowledge and confidence in the positive approach is essential to success in pistol shooting but the ever present danger of the adverse effect of the negative thinking makes it necessary to know the characteristics of that inviting trap.

    C. <u>Third Student Performance Objective</u>: The pistol shooter-student should know why negative thinking can prevent a competitor from winning.   (5 Min)

    <u>QUESTION</u>: Why is it so difficult to shoot championship scores?

    1. It's not that most of us have not been taught the fundamentals of pistol shooting. The fault usually lies in that we open our minds up to thousands of negative reasons why we cannot shoot good scores.

        a. Bad weather, rain, cold, sun, wind, etc.

        b. Inefficient range operations.

      c. Below standard equipment and ammunition.

      d. Lack of incentive.

      e. Competion too tough.

      f. Afraid to win.

      g. Carelessness.

      h. Overconfidence.

      i. Pessimism.

      j. Exposure to distractions.

There are probably numerous other factors but these are a few that were pointed out by top level shooters. We know that we must exclude factors that detract from good performance and use those remaining factors to our advantage.

QUESTION: Who won the last match in which you participated? To achieve results on a level that will produce winning scores in today's competition, it is necessary to have a coordinated, exacting control of the technique of employment of the fundamentals based on the capability for intense concentration. Each properly executed sequence of actions that creates conditions for a good shot, contributes to the ease with which it can be repeated.

TRANSITION: If you think that you and you alone have the problem of match pressure, look around; we all have it. The man who has never experienced match pressure has never been in a position to win a match. Where is the difference? Where is the dividing line between champion and duffer? Both may shoot comparable scores in practice, yet one is invariably at the top of the bulletin and the other on the second page. The dividing line is clear and obvious; the ability or lack of control in their thinking. Mental discipline! Some shooters have learned to control their emotions and anxieties and go right ahead and perform within their capabilities. Others even with years of experience, and also with a wealth of doubts and negative thoughts, pressure themselves out of the competition every time they step up to the firing line.

    D. Fourth Student Performance Objective: The pistol shooter-student should have an understanding of the causes, effects and control of match pressure.   (10 Min)

      QUESTION: What causes match pressure?

      1. First in our treatment of match pressure we must find what causes it, for without knowing the precipitating factors we can never combat it. Match pressure is the direct result of the fear of failure and the loss of self esteem. Are we afraid of winning? If this were the real cause, we would have no desire to win, or to perform well, and there would be no pressure. No, it is not the actual winning we are afraid of. We are afraid of not winning. We are prone to succumbing to our fear of performing poorly and having our fellow competitors see our poor performance.

      2. What happens to us physically when we are subjected to all of these mental gymnastics that result from match pressure. First and most prominent, we shake, we drop our magazines, put our scope on the wrong target and some of us even shoot on the wrong target. In short we commit what seems a series of asinine mistakes that normally would never occur; the type of things that result, from the preoccupation of our minds with, in this case,

useless psychological distractions. In addition you invariably experience a shortness of breath which increases your breathing rate, and your heart beats about twice as fast as necessary. All of which seems to make it impossible for us to hold our pistol reasonably steady, let alone shoot well. To add to our distress we feel that everyone is witnessing our anxiety and stupidity. Yet with all this, our counterpart the champion appears to be calm and enjoying himself. Let's face it. He is!

3. There are definite advantages to match pressure. Many of our senses are more acute. For our purposes we see better, and our sense of touch is more exacting (that is why your trigger seems to become heavier in a match; actually it has not changed a bit, but we are more aware of it). Our awareness of the passage of time becomes more vivid. Don't believe it? What about the anxiety you feel just before you shoot the last round of a rapid fire string. All of these added together should, if employed correctly, make us more exacting and consequently better our performances.

4. How do we control match pressure? First, realize that it can be controlled and actually used to your advantage.

    a. Prior Mental Determination. This is the greatest asset that we have available to us. By thinking through the correct procedure for firing each shot just before you shoot, and making up your mind to do it the correct way, you can virtually eliminate distractions in the actual execution. Be warned right now, that if you fail to do this and approach the shot without a preconceived plan of attack, or without the mental determination to be right come fire or flood, your results at best will be erratic. You readily appreciate the necessity for concentrating on and aligning your sights. A very effective way to assist in this is to sit down and close your eyes and imagine front and rear sights including the blurred target. Try it right now. Most of us find that it is almost impossible to keep them aligned perfectly even in your mind's eye. However, by doing this, you are conditioning your mind to be able to focus the mental concentration where you want it to point. As a result, it becomes that much easier for you to do it on the range. This technique of mentally aligning the sights is very effective if practiced just before attempting to fire a controlled shot.

    b. Channel Your Thinking to the More Important Fundamentals. You must continually think fundamentals and review them in your mind. Train yourself so that as many as possible of these fundamentals are executed automatically without any tedious effort on your part. When you do this, it leaves you with only the most prominent fundamentals to contend with in the actual firing, sight alignment and trigger control. This will enable you to, as an example place all of your mental and physical efforts toward keeping the sights aligned and smoothly releasing the hammer while your position and arc of movement are so well ingrained as a result of training that you will employ them correctly automatically.

    c. Establish a Routine. From routine comes boredom. What is boredom? The lack of excitement. What are we trying to do? Keep from becoming excited. In a more serious vein, however, in establishing a routine, you eliminate the possibility of forgetting some trivial item of preparation that may throw you off balance later if you neglect it.

    d. Work on Each Shot Individually. Or, in the case of timed or rapid, each string of five shots. Each shot must be treated this way for in reality there is no reason to believe that because your first shot was an eight your next one will be the same. Nor is it logical that if your first three shots were tens, you have a guarantee that those to follow will also be tens. Each one is merely a representation of your immediate present ability to apply the fundamentals correctly or incorrectly. And your ability to do this will vary considerably if you let it. Do not connect the shot you are preparing to fire with the value of those already on the target. The performance requirement demanded of you to control this shot to be fired now is not dependent on the value of the previous shot or on the value of shot to be fired immediately after it.

e. Win the Aggregate, Not Just One Match. Why should we become excited or worried when we have cleaned three of the four strings of the 45 cal timed fire match. Go right ahead and clean the next string. Sure, if you do so, you may win the 45 cal timed fire match, but that is not your overall objective. You came here to win the aggregate, not **just one** match. Don't drop a couple of points here just because the possibility of winning one match has arisen.

f. Train Yourself to Think Performance Rather Than Score. Employing this technique, an eight or a seven becomes not a shot that subtracts two or three points from your aggregate, but a shot where you allowed yourself to deviate from proper employment of one of the fundamentals. Rest assured that if you do your part on the firing line, the score will take care of itself.

g. Who Said "Stay Out Of the Scope?" If you are shooting a slow fire match you must go down and score after ten shots, and if it's National Match Course, you must score the slow fire stage before you shoot at 25 yards. Do you think you are going to keep something from yourself? Why should a good score scare you? A good score is just exactly what you went up to the firing line to accomplish. Of what value is a 98 slow fire if you don't possess the fortitude to continue a good performance? Learn to use the scope for the purpose it was intended. A check on your performance and zero. Use your scope as an aid in your analytical procedure, not to score your target. We are not so pretentious that we believe we are going to go through a ten shot string slow fire with only three nines and not know what our score is. Our scope, is, once again, to be used to evaluate the end product of our performance.

h. Relax Your Mind, right from the time you get up in the morning. Nothing will put you in a greater state of mental agitation than to rush through breakfast, rush to get to the range just in time to make your relay. If this happens, your slow fire is ruined at about the third red light you hit. Take it easy. Shooting is fun, enjoy it.

i. Practice Tranquility. Ever see the guy that loses his temper every time he has a bad shot? Who is he mad at? Those individuals who lose their temper are doing nothing more than exhibiting self-admonishment for their vacillation in the execution of a shot. They recognize that if they had worked a little harder applying the control factors, the shot would have been better. On the other hand if we do everything within our power to make the shot good and for some reason or other it isn't good, we should have no cause for undue irritation. Although a good shooter must exert all of his mental and physical ability toward shooting a good score, infrequently he will fail to do this. Suffice to say that when this happens, if he chastises himself severely, or falls into a fit of complete depression because of a poor score, he will hurt greatly his chance for the rest of the match. It is not intended that you laugh off or treat lightly a poor performance; however, you must possess the presence of mind to accept the bitter with the sweet. Preparation, planning, relaxing, delivery of the shot, careful analysis and positive corrective measures, is the cycle of action you must force yourself to conform to without deviation. You can then be assured that the next shot can be delivered under the most precise control you are capable of exerting at the present moment.

j. Match Experience. Without question, competitive experience is one of the ingredients necessary in the making of an accomplished pistol competitor. However, experience alone is of no value. We must flavor our experience with an accurate and honest evaluation of our performance and the positive corrective measures that will raise our ability and eventually our scores. We must experience an increasing degree of mental control. It is not easy and is often left out of our training until our physical ability to shoot far outreaches our ability to exercise control when the chips are down. Perhaps when we first hand a youngster a pistol we should say "These sights are the two things that you must train your mind as well as your hand and eye to control". Instead, we usually point out the pretty cow horn stocks and shiny barrel.

k. Physical Conditioning. There is no doubt whatsoever that you can shoot better if you are in good physical condition. Your ability to hold, for example, is no better than the ability of the muscles of your arm to do this for you. Your ability to resist the stress and strain of match pressure and anxiety is directly in proportion to your physical condition.

l. Argue With Your Subconscious. Not only argue with it but win the argument. Even as we read this some of us are hearing that little voice in the back of our minds that keeps saying "Yes, this sort of thing may work for Joe, but I know damn well I'm going to goof up the next time I get close to a winning score." Who's voice is this? Where did all these ideas come from in the first place? Where did this little guy get all his knowledge? Let's be realistic. Our conscious mind puts these ideas into our subconscious, so don't ever believe that you can't overpower it. It's not easy. He's been saying what he pleased for years and now he isn't going to be routed easily. But don't give in to him and eventually you will find that the subconscious mind is not in conflict with your conscious efforts.

m. Now with all of this emphasis on the positive approach you are going to get a big "don't."

(1) Don't expect immediate results the first time you try mental discipline. There is a coordination of employment of the fundamentals to contend with and first, you must master the control of these. There are no hidden secrets. All that we gain is the direct result of hard work. If you find that you can exercise satisfactory control only for a short period of time, work on extending this period by practicing and perfecting your system. Remember that your returns are in proportion to your investments.

TRANSITION: The fear of failure to perform up to your known capability will generate gradually increasing tension.

E. Fifth Student Performance Objective: The pistol shooter-student should know how to reduce the effect of tension and attain relaxation under conditions of stress.

QUESTION: What are the different types of tension?

1. Types of Tension. Normal tension is a blessing to mankind. Without tension most problems could not be solved; the world's work would not get done and championship scores would not be fired. Normal tension is the prevailing condition of any organism when it is mustering its strength to cope with a difficult situation. All animals, including man, tense in situations which involve the security of themselves and their dependents. But there is a kind of tension that is bad for you: pathological tension. This is an exaggeration of normal tension, and thank heaven, fairly rare. The vast majority of people who worry about it have nothing more than normal tension. All they need is a technique for relaxing. We should know what tension really is and here are a few hints on how to terminate it.

2. Tension Reducing Techniques:

a. Take a Breather. Breathe deeply, three times, very slowly. At the end of each exhalation hold your breath as long as possible. When you have finished you should feel noticeably relaxed and much calmer.

Here's what has happened. By forcing yourself to breathe deeply you break the tension of your voluntary breathing muscles causing the involuntary muscles of the lungs, gastro-intestinal tract and heart to relax too. This is the simplest method for relaxing. For some, it can be used to end tension completely. It can be used by others for temporary relief when they do not wish to "let down" completely.

b. Let Go! Sit down and let your head droop forward. In about a minute raise one arm and drop it in your lap as if it were a limp rag. Do the same with the other. Now let your legs go completely limp; now your stomach muscles. Stay in this position for at least ten minutes. This technique, too, is aimed at first relaxing the voluntary muscles. It is especially effective when you've had to maintain normal tension for several hours on end.

c. Shift into low. When you have been overstimulated by highly demanding mental exertion, taper off by becoming involved in a diverting activity. If you like handiwork, pick a kind which interests you but is not too creative. Soap sculpture, finger-painting, woodworking, and gardening all are excellent low-gear activities that will help you to simmer down. This kind of tension-remover is aimed at changing your mental set. It is helpful for those who have to operate at top capacity such as shooters and are in enforced contact with others all day long. After stimulation, a part of you wishes to continue to be diverted. To slow you down when you're in this state of mind, you require something which is engrossing but which demands nothing of you intellectually. Television suspense plots and simple handicrafts are ideal.

d. Take a break. This is a "remote control" technique for dealing with normal tension. Simply take a break for ten full minutes every hour. You may find that this allows you to ease out of your working tension more quickly and easily when the day is over. The reason this works: since you have not allowed tension to develop fully, your organism doesn't, so to speak, have so far to go on the road back to normal relaxation.

e. Stop and think. When the tension-making job allows a respite, sit down and calmly review the things in your life that you value highly. Think of the long range purpose of your life, of the people you love, the things you really want. In a few minutes you may notice that you have involuntarily taken a deep breath. This is a sign that tension is dropping away rapidly. When you tense to face a difficult situation, you tend to exaggerate its importance. Judgment and reason can quickly change this mental state when it's time to relax again.

These techniques are based on the fact that tension can be ended in two distinct ways: through the relaxation of your voluntary and involuntary muscles; and by changing your mental "set". If you achieve either, you set off the other and hasten the process of normal relaxation.

VI: CONCLUSION: (5 Min)

A. <u>Retain Attention</u>: Be a hungry shooter. The slashing, no holds barred drive for victory in competition, from the first shot on, destroys the confidence of the lesser competitors

B. <u>Summarize</u>:

1. Mental discipline is essential for advanced pistol shooters.

2. Development of mental discipline, confidence and the ability to think positively.

3. Why can't you be a winner? The dangers of negative thinking.

4. The causes, effects and control of match pressure.

5. Reducing the effects of tension and attaining relaxation under conditions of stress.

C. <u>Application</u>: The self control attained by the advanced pistol shooter pays off not only in better match scores, but also in combat situations, by the calmness and resolution in using his weapons to kill with nearly every round fired.

D. **Closing Statement:** Mental discipline provides the emotional stability so necessary to the development of the champion shooter. Confidence in his ability and mastery of the basic skills combine to produce a dependable performance under all degrees of stress.

PREPARATION DATA SECTION

I.	TITLE AND NUMBER:	Mental Discipline (P-8)
II.	TIME ALLOTTED:	Fifty (50) minutes
III.	PERIODS OF INSTRUCTION:	One
IV:	CLASS PRESENTED TO:	Coaches and Instructor-Shooters
V.	INSTRUCTOR REFERENCES:	The Advanced Pistol Marksmanship Manual
VI.	INSTRUCTOR AIDS:	1 Podium, Podium Cards, 1 sound Set, Chalk, Blackboard, GTA Charts of 35mm slides and Projector
VII.	STUDENT EQUIPMENT AND UNIFORM:	Uniform as Prescribed
VIII.	PHYSICAL FACILITIES:	Outdoor Bleachers or Classroom for Students
IX.	PERSONNEL REQUIREMENTS:	1 Principal Instructor, 1 Asst Instructor, 1 Sound Set Operator.

COACHING I

ATTRIBUTES, RESPONSIBILITIES AND DUTIES
OF A PISTOL TEAM COACH (P-9)

LESSON OUTLINE

I. LESSON OBJECTIVE: To enable small arm firing school students to understand the attributes, duties, and responsibilities of coaching a pistol team.

II. STUDENT PERFORMANCE OBJECTIVE: As a result of this instruction the student must be able to:

    A. UNDERSTAND the personal attributes of the pistol team coach.

    B. EXPLAIN the head coach's responsibilities.

    C. DEFINE and EXPLAIN the line coaches' duties.

III. ADVANCED ASSIGNMENT: Advanced Pistol Marksmanship Manual, Chapter IX, "Attributes, Responsibilities and Duties of a Pistol Team Coach."

IV. INTRODUCTION: (3 Min)

    A. Gain Attention: (Have a student ask the question) "Where were you last night, coach?" (Answer) "At home in bed with my wife, sober and asleep. Where else would a good coach be the night before a match?"

    B. Orient Students:

        1. Lesson Tie-In: You have been previously instructed in the fundamentals and techniques of shooting a pistol. We will now discuss the qualities and requirements necessary to properly guide a pistol team.

        2. Motivation: All of you probably will at one time or another find yourselves acting as a line coach, possibly not during a team match, but at least coaching another shooter.

        3. Scope: During this period we will discuss the attributes and responsibilities of a pistol team coach and his duties before, during and after a match.

V. BODY:

    A. First Student Performance Objective: Students must understand the personal attributes of the pistol team coach. (12 Min)

        QUESTION: Name some of the attributes of a pistol team coach.

    A coach must be strict in his demands upon his shooters and consistent in what he requires. He must insist always on observing discipline and on adhering to the program. There are attributes, moral and mental, that the pistol coach must have that will accelerate the shooter's progress and prevent a lapse into habits that may lead to a decline in performance.

        1. A coach must be temperate in all things. Nothing should preclude the attainment of what he knows to be his prime objective. Win.

2. The coach must be <u>dedicated</u>, but it is not necessary that he be a champion shooter. Many hours are spent by the champion in perfecting his own skills. The demand on his working hours leaves him little time to spend coaching and helping others. However a good coach will devote most of his time helping others even if he must allow his own scores to suffer. However, this does not mean that the champion shooter would not make a good coach, but the time spent away from his work is costly to him if he wants to continue to be a champion.

3. <u>Self control</u> and an infinite amount of patience are two main attributes of a pistol coach. Many times during team matches, under pressure, things could be done and said which would have an adverse effect on the team. The ability to control himself and the other members of the team will affect greatly the outcome of the team match.

4. The coach must be <u>compatible</u>. If the coach cannot get along with his shooters, he is worse than useless to the team.

5. You as a coach, must be able to <u>inspire confidence</u> in your shooters. First you must have confidence in yourself and show your team members that you have complete confidence in them and their ability, both as individuals and as team shooters. Then and only then, can you expect them to have confidence in their abilities and just as important, confidence in you.

6. The coach must show <u>enthusiasm</u>, not just when the team is winning. If the team has lost a match, they must rekindle a desire to win and feel a source of enthusiasm to be able to come back and shoot better in the next team match. Desire to win and enthusiasm are the things that keep us from giving up in disgust when everything seems to be going wrong.

7. The pistol coach must be <u>observant</u> of anything in the shooter's performance that can be improved to make his scores better. This applies primarily in practice, when the time can be taken to diligently pursue improvement, but also should be observed in matches. A close check on all phases of the match such as scoring, alibis, challenges, and range officer decisions must be made by the coach.

8. The pistol coach must have a <u>wide knowledge</u> of the shooting fundamentals and coaching techniques. In addition he must have <u>specific knowledge</u> of match rules and range procedures. The coach must constantly keep in his mind the traits of each shooter. He must know how to handle each one in every shooting situation so that the utmost can be obtained.

9. You must set <u>exacting standards</u> for the shooters to follow. Your actions, as a coach, and as a member of the team must be above reproach at all times, you must require your team to utilize practice sessions to improve their shooting and to try new ideas. During the shooting, the coach should require a great amount of effort devoted by each man toward correcting his faults, and not allow him to accept an average performance. Strive for perfection in shooting, but at the same time retain exacting standards of personal and team conduct.

10. The coach must have an <u>open, progressive mind.</u> Be able to accept constructive advice from anyone. No two individuals are alike. The coach who is flexible in his dealings with his shooters will in the long run have the best functioning team.

<u>TRANSITION:</u> That briefly is some of the personal attributes of the pistol team coach. As headcoach of a complete pistol division he is responsible for the success of a whole group of teams.

B. <u>Second Student Performance Objective:</u> Students must be able to explain the head coaches' responsibilities. (15 Min)

<u>QUESTION:</u> How does the head coach know what training is necessary?

1. <u>To determine training requirements</u>, a coach should determine the shooting ability of each of his shooters. This ability of course varies greatly not only between All-Army level and the Camp, Post, or Station level but, of more interest to the coach, between the individual shooters on the same squad. This is a difficult but very important job. What the coach selects as the needed training subjects may well determine the later success of the team.

2. <u>Publication of a training schedule will not suffice to insure progress in training</u>. Careful observation of individual members and correcting faults during the training is one of the primary duties of the coach. The coach will initially use the scores fired in practice to assist in selecting team members and future scheduling of training. Later only match scores will be used. See Chapter XI, "Evaluation", Volume I.

3. The coach should <u>periodically check the equipment</u> of each SHOOTER. This check serves a quadruple purpose.

    a. Proper mechanical functioning of weapons.

    b. The inherent accuracy of each weapon.

    c. Cleanliness of weapons.

    d. Regular serial number checks of weapons issued to individual shooters and security storage inspection insures proper accountability and security of all weapons and equipment.

4. The coach should <u>conduct periodic written or oral tests</u> in order to check how much the shooters have learned. This means how much of the fundamentals, of the NRA Rules, of general match procedure and how well they have mastered techniques. The results should become part of the shooter's record.

5. Constantly maintain a <u>current evaluation of each shooter</u>. An analysis of the rate of progress, individual morale and attitudes, the degree of team effort exercised, all of which should be current and decisive.

6. <u>Establish current doctrine and performance standards</u> by constant review of training manuals, training materials and methods so as to reflect new ideas and methods proven to be sound and reliable and weed out unsound techniques.

7. <u>Improve the team potential</u> by conducting periodic, organized, group instruction.

8. <u>Improve the individual shooter's potential</u> by personal and private interviews and conducting individual coaching sessions.

9. <u>Supervise the team preparation</u> for match participation.

10. <u>Supervise the coaching technique</u> of individual line coaches.

11. <u>Assists in the preparation of instructor courses</u>.

12. <u>Assists in rehearsal of instructors</u>.

13. <u>Participates in all competition</u>. A shooting coach is a person who appreciates the great problems faced in competition and is not inclined to be arbitrary in his judgement.

14. Exercise a profound influence on the morale, attitudes, enthusiasm and the shooters' will to win by exhibiting individual consideration, stimulation of confidence and creation of an atmosphere of inevitable success. Pride and esprit de corps go hand in hand with an attitude that reflects the will to win.

15. Make corrections and recommendations. During the match, the head coach should keep notes on any mistakes made by a team member, or the team. Then put into his training program the material necessary to bring about corrections. If he fails; the team fails, if he is a success the team will be a success.

TRANSITION: The successful exercise of the responsibilities of the head coach go hand in hand with the careful performance of duties by the line coach.

C. Third Student Performance Objective: Students must be able to define and explain the line coach's duties. (15 Min)

QUESTION: What can be done before that all important team match to make sure that nothing is overlooked that will jeopardize your teams' chances of winning?

1. Make your self available during the firing of the individual matches. You should check the performance of your team members and, if they are falling below standard, talk to the shooter concerned between matches and try to find out the nature of the trouble. Make yourself available to your shooters to assist them in every way possible.

2. Supervise physical preparation. After the individual matches and prior to the team matches the coach should have a team meeting. During this meeting he should insure that all the team members have the right type and amount, of ammunition. Are their weapons clean and in perfect working condition? Check everything. At this time assign relay and target numbers to team members.

3. Supervise mental preparation, refer to coach's worksheet.

4. Make the decisions for your team. During the match, your team is on the firing line. What are your actions? You, as the coach, must give positive directions, tell them what to do. You are responsible for the team. You are the person who must decide what is to be done and how to do it.

5. The coach must be constantly aware of any breaches of safety.

6. The coach must cater to the shooter's needs when on the firing line. The red carpet treatment breeds respect and trust.

7. You must encourage the members of your team at all times. When they are shooting good you must encourage them to shoot better. When they don't do so well, try to assist them in mental preparation for shooting the next match. Never say anything to discourage any member of your team. By the same token don't use false praise.

8. All scoring of practice matches and of record matches must be supervised by the coaches. This doesn't mean that every shot fired must be counted for score. The shooter must be given free practice to allow them to try new procedures such as different positions, grip, etc.

9. After the match, critique your team. The coach and team should go over their performance and suggest ways to improve this shooter's rapid fire or that shooter's slow fire, etc.

10. The coach must constantly evaluate and observe the members of his team. Direct coaching of the shooter is only half of the job. The line coach will evaluate the shooter's potential ability, since he will be in the best position to see how the shooter reacts to the problems he is confronted with when trying to achieve a good score.

VI. CONCLUSION:

A. Retain Attention: What is the mark of the effective coach? A halo or a scope, a manual and a gun box?

B. Summarize:

1. Personal attributes of a pistol team coach.

2. Responsibilities of the pistol team head coach.

3. The line coach's duties.

C. Application: The results achieved will be far greater by having coaches that are fully qualified to control the performance of the shooters on a team that is expected to win or set a new record.

D. Closing Statement: Behind every good shooter or winning team is a good coach and a lot of hard, well planned work. Hope or luck will never win the match.

PREPARATION DATA SECTION

I.	TITLE AND NUMBER:	Attributes, responsibilities and duties of a pistol team coach (P-9)
II.	TIME ALLOTTED:	Fifty (50) minutes
III.	PERIODS OF INSTRUCTION:	One (1)
IV.	CLASS PRESENTED TO:	Coaches and Instructor-Shooters
V.	INSTRUCTIONAL REFERENCES:	Advanced Pistol Marksmanship Manual.
VI.	INSTRUCTIONAL AIDS:	Podium, podium cards, blackboard, Chalk, GTA Charts or 35mm Slides & Projector, 1 sound set.
VII.	STUDENT EQUIPMENT AND UNIFORM:	Uniform as prescribed.
VIII.	PHYSICAL FACILITIES:	Bleachers or classroom.
IX.	PERSONNEL REQUIREMENTS:	1 principal instructor, 1 assistant instructor, 1 sound set operator.

COACHING II

TECHNIQUE OF COACHING A PISTOL TEAM (P-10)

LESSON OUTLINE

I. LESSON OBJECTIVE: Familiarize the coach and shooter with the techniques of coaching a pistol team.

II. STUDENT PERFORMANCE OBJECTIVES: As a result of this instruction the student must be able to:

A. UNDERSTAND the intricate tactics and consummate skill involved in coaching a champion pistol shooter.

B. COMPREHEND the importance of using a systematic coaching technique.

C. EXPLAIN in general terms, the technique used by the head coach in training and handling an entire pistol team division.

D. ORGANIZE and control a four-man pistol team in the assembly area and on the firing line.

E. ANALYZE present coaching technique with the purpose of incorporating improvements that will be reflected in team members' shooting performance.

F. CONDUCT a demonstration of pistol coaching technique.

III. ADVANCE ASSIGNMENT: Advance Pistol Marksmanship Manual, Chapter X, "Technique of Coaching a Pistol Team."

IV. INTRODUCTION: (3 Min)

A. Gain Attention:

NOTE: HAVE AN ASSISTANT INSTRUCTOR FIRE FIVE (5) BLANK ROUNDS OF RAPID FIRE WITH A 38 CAL REVOLVER AT A 25 YARD STANDARD AMERICAN PISTOL TARGET. THE PRINCIPAL INSTRUCTOR STANDS BESIDE THE SHOOTER AND ADDRESSES HIM AT THE COMPLETION OF THE STRING OF FIVE SHOTS. A BRIEF DESCRIPTION OF THE SIGHT ALIGNMENT OF EACH SHOT AND CALLING OF THE SHOT GROUP LOCATION IS REQUESTED. THE INSTRUCTOR VIEWS THE SHOT GROUP THROUGH THE SPOTTING SCOPE AND COMPARES THE CALL AND SHOT GROUP. IT IS SATISFACTORY AND A STATEMENT TO THAT EFFECT IS VOICED TO THE SHOOTER.

The probability that the pistol team shooter will have a consistently controlled performance is immeasurably enhanced by systematic guidance and helpful analysis of his shooting by a qualified coach.

B. Orient Students:

1. Lesson Tie-In: This hour of instruction is the second of a three part block of instruction on coaching. During the first hour you considered the necessary attributes, responsibilities and duties of a pistol coach. This period will be spent discussing the techniques of coaching that result in aiding the control of performance by the shooter.

2. _Motivation_: Many of you will probably one day find yourselves acting as a pistol team line coach or at least advising and coaching another shooter with the intent of helping him to improve his pistol shooting performance. What will be your method of operation?

3. _Scope_: You learn to coach a pistol team or a shooter effectively by knowing:

    a. The intricate tactics and consummate skill involved in guiding and advising the champions.

    b. A systematic coaching technique.

    c. The technique the head coach uses in training and handling a pistol group composed of multiple four man teams.

    d. How to organize and control a single four man pistol team.

    e. A method of analysis that will reveal shortcomings in coaching technique.

    f. How to demonstrate coaching methods.

NOTE: ASK IF THERE ARE ANY QUESTIONS ON THE ADVANCE ASSIGNMENT.

V. BODY:

    A. _First Student Performance Objective_: The coach must understand the intricate tactics and the consummate skill involved in coaching a champion pistol shooter. (10 Min)

       QUESTION: Intelligent coaching will or will not help the champion pistol shooter?

       1. The champion is a superb performer and is capable of unexcelled accomplishment but he invariably suffers lapses of poor control. The guidance of a good coach will help to restore his control and the champion continues toward his objective: a record breaking performance.

       2. The development of a shooter must have direction even though continuous, stable progress and performance is greatly dependent upon his determination to become a champion.

       3. The shooter's ideas and techniques are profoundly influenced by personal coaching.

       4. The delicate ego, the vulnerable confidence of a shooter suffers immeasurably from an unexpected or unresolved error of performance.

       5. An important part of the coaches' function is to sustain confidence and remove any cause for self-doubt on the part of the shooter.

       6. A good shooter's self-respect needs tender care.

       7. A good coach can improve the shooters performance.

       8. The formula that the pistol coach should use to bring forth the winning results is basically motivated by mutual confidence and trust.

TRANSITION: An understanding of the complexities of counseling a champion is put to use efficiently by the systematic employment of the factors of proper pistol coaching.

B. **Second Student Performance Objective:** The factors of systematic coaching technique are six in number. (5 Min)

QUESTION: What are the factors of systematic coaching technique?

1. Physical and mental preparation must be carefully and completely performed.

2. The sequence of events, actions and thought processes necessary to the delivery of a controlled accurate shot requires planning in great detail.

3. The shooter must achieve relaxation of mind and muscle in face of stress.

4. Deliver the shot on target, exercising control as planned.

5. Make a comprehensive analysis of the errors, if any, in the delivery of the shot.

6. Positive corrective measures should be incorporated into the planning of the next and subsequent shots in an effort to prevent the recurrence of the error.

TRANSITION: After gaining a command of general coaching methods, it is paramount that the coach attain a grasp of the larger mission: As a Head Coach, directing the actions and training of a group of pistol teams.

C. **Third Student Performance Objective:** The Head Coach of a Pistol Team must have a technique for handling and training a group of pistol teams. (10 Min)

QUESTION: If a satisfactory method of coaching is developed, should all shooters training under a coach be required to "Hew to the Line"?

1. Every competitive shooter must be handled with an individual approach commensurate with his peculiar needs.

2. Training methods should be flexible enough to allow work on specific difficulties experienced by individual competitors.

3. Study shooters behavior during competitive match shooting.

4. Evaluation of the team shooter is based on direct comparison of attributes and performance with team mates and competitors.

5. The accomplished marksman enjoys a sense of ingrained discipline for excellence and is motivated to exceed the challenge of existing standards of victory.

6. Recorded practice is limited to half of total practice time.

7. A high level of morale and espirit de corps must be maintained if a winning performance is expected.

8. The team shooter must not be overburdened with a fear that his performance, if below average, will jeapordize the team's chance of winning.

9. The training period before a match should duplicate the program and conditions of the impending match.

10. The team members should be allowed one or two days rest immediately prior to participation in match competition.

TRANSITION: All shooters are not champions and many promising aspirants need close supervision. The line coach is in intimate contact with his four man team during long hours of training and in the heat of team competition.

    D. Fourth Student Performance Objective: The line coach of a single four man pistol team must have a technique for organizing, controlling and training the individual shooters in his group in the assembly area and on the firing line. (7 Min)

    QUESTION: When a shooter gets nervous in team competition, how does a coach calm him down?

    1. The shooter's nervousness usually stems from fear of failure. Reassurance by reorientation in the technique of fire will usually lessen anxiety and restore control.

    2. A coach must know a shooter's weaknesses, strong points, peculiarities, etc. to avoid unintentional harmful effects on shooter's performance.

    3. A coach must control the tempo of shooting by the individual team members.

    4. A coach can assist the shooter to reestablish coordination when control wavers.

    5. A coach must cope with an ever changing situation in the shooter and the match conditions.

    6. A coach must not overestimate the benefits of dry firing.

    7. The coach must say the right word at the right time.

    8. A friendly, cooperative attitude will help the team members to shoot with assurance and boldness.

    9. The coach must decide the order and pairing of relays by compatibility and fortitude characteristics of team members.

    10. Team selection is based on a standardized evaluation process and when a decision is reached, it should remain firm.

    11. The developement of a shooter depends upon diligent, creative work by the shooter and coach.

    12. A shooter must develop the ability, through systematic analysis to determine what factors are affecting the accuracy of his shooting.

TRANSITION: A good technique for this shooting season may be found wanting next season.

    E. Fifth Student Performance Objective: Analyze present coaching technique with the purpose of incorporating improvements that will be reflected in team members' shooting performance.

    QUESTION: If a shooter doesn't want to talk between shots or strings how does the coach communicate with him?

1. Ask questions, conversation furnishes the clues to performance that are not apparent in observation.

   2. The coach or shooter should not expect spectacular improvement as a result of a few well chosen words or corrective action.

   3. Why some shooters do not shoot well.

   4. Why some shooters do shoot well.

   5. Teams and coaches on the same squad must cooperate by exchange of helpful information.

   6. A good coach knows when it is important that he not coach.

   7. Good results are achieved by hard work.

   8. The coach is in charge of the team.

   9. Do not try to correct too many mistakes in one training session.

   10. If team shooters desire to coach each other, all the established functions of coaching must be adhered to.

   11. The best coach-shooter relationship is a mutual understanding which creates conditions for producing the highest scores possible.

TRANSITION: All the established factors of pistol coaching technique become actively significant when a working example is observed.

   F. Sixth Student Performance Objective: The conduct of a demonstration will amplify the factors important to establishing effective pistol coaching technique.   (10 Min)

   1. The preparation stages are important insurance that nothing will be overlooked that can help win the match.

      a. Preparation in the assembly area.

NOTE: POINT OUT TO THE CLASS AND COMMENT UPON PREPARATIONS BEING SUPERVISED BY THE LINE COACH.

"FIRST RELAY - .45 CAL TEAM MATCH TO THE FIRING LINE"
            (See Figure 1, Chapter X for organization and placement of a pistol team.)

      b. Preparation (3 min) on Firing Line (Shooters slowfire worksheet - page 94) (Chapter IV, Volume I)

         (1) Physical)
)
) - (Coaches worksheet page 164 para. a and b)
)
         (2) Mental)

         (3) Check stance and position for natural hold.

(4) Check grip for natural sight alignment.

2. <u>Planning Shot</u>: (Review shot sequence-slow fire worksheet page 94)

<u>FIRE COMMAND - "LOAD"</u>

3. <u>Relaxation and Breath Control</u>. (Deep breathing before shooting)

   a. Relax all muscular systems not necessary for stable, upright posture.

<u>FIRE COMMANDS TO BEGIN SLOW FIRE - "IS THE FIRING LINE READY?" ETC.</u>

4. <u>Deliver Shots as Planned</u>. (Do not compromise)

5. <u>Shot Analysis</u>. (Shooter's Remarks)

   a. First Shot: Fired too quickly (Breakdown in follow through - trigger pressure applied abruptly, shot call unsure, sight alignment excellent before trigger pressure was applied)

   b. Second Shot: No shot call (No clear impression of sight alignment.

   c. Third Shot: Called good after one unsuccessful try. Benched weapon (No error in sight alignment or hold)

   d. Fourth Shot: Holding too long (Shot broke as arm moved toward two o'clock; gross errors corrected before applying trigger pressure)

   e. Fifth Shot: Called good (Describe sight alignment, shot is low ten. Arc of movement was in lower half of aiming area)

6. <u>Positive Correction:</u> (Coaches Remarks)

   a. First Shot: Probably applied trigger pressure on basis of sight picture. Unchanged rate of positive trigger pressure will insure surprise break of shot. Once trigger pressure is started, you are committed to continue at the same rate until shot breaks.

   b. Second Shot: Focus allowed to move to target momentarily and shot was fired during this short interval. A prior determination to pinpoint focus on front sight until shot breaks will avoid this lapse.

   c. Third Shot: If error is so great as to cause let up in trigger pressure before shot breaks, bench weapon, analyze trouble, correct error, replan delivery of shot and try again. In planning for the next shot, remember the exact sequence in which the factors for controlling a good shot were applied.

   d. Fourth Shot: Holding too long will cause impatience and creates a tendency to speed up trigger pressure rate of application. Anticipation of the shot breaking will cause a somewhat less violent reflex in the form of a straight line arm movement rather than the muscular twitch that usually follows a more abrupt increase in trigger pressure. If error is so great as to cause let up in trigger pressure before shot breaks, bench weapon, analyze trouble, correct error replan delivery of shot, relay and try again.

e. Fifth Shot: Normal in all respects. Arc of movement, however small, is present in the ability to hold of all shooters. The law of averages will have some shots break near the edge of your ability to hold. Maintainance of good sight alignment while waiting for shot to break will prevent wild shots.

FIRE COMMAND - "CEASE FIRING"

FIRE COMMAND - "CLEAR WEAPONS"

"SCORE AND REPLACE WITH A 25 YD TARGET"

7. <u>Overall Critique</u> of slow fire while targets are being scored.

8. <u>Physically and Mentally prepare for rapid Fire</u>: (Shooters rapid fire worksheet page 95) (Chapter IV - Volume I)

    a. Physical (Set sights for 25 yds)

       (1) Check position for natural hold, etc.

"YOU MAY HANDLE YOUR WEAPONS"

       (2) Check grip for natural sight alignment, etc.

    b. Mental (Coaches worksheet - page 164 par b)

9. <u>Plan 5 shot string of rapid fire</u>: (Review shot sequence - rapid fire worksheet page 95).

FIRE COMMAND - "LOAD 5 ROUNDS"

10. <u>Relaxation and Breath Control</u>: (Deep breathing before shooting)

    a. Relax all muscular systems not necessary for maintaining upright posture.

FIRE COMMANDS FOR RAPID FIRE - "IS THE FIRING LINE READY?" ETC.

11. <u>Deliver 5 Shots as Planned</u>: (Do not interrupt rhythm to make corrections in sight alignment and hold. Grip and position check insures a recovery from recoil that will have few errors that need correction.)

12. <u>Shot Analysis of 5 Shot String</u>: (Shooter's Remarks)

    a. First Shot: Late. (Looking at target on turn. Focus would not return quickly to sight alignment.)

    b. Second Shot: Fired too quickly. (Time anxiety caused speed up of trigger pressure; front sight dipped.)

    c. Third Shot: Fired normally after settling into minimum arc of movement, picking up sight alignment and pressing positively on trigger.

    d. Fourth Shot: Same.

  e. Fifth Shot: Same.

 13. Positive Correction: (Coach's Remarks)

  a. First Shot: Concentrate point focus on front sight and be aware of proper alignment with rear sight notch. When the targets begin to face, shooter should start application of positive trigger pressure.

  b. Second Shot: Straight through, positive, unchanging rate of trigger pressure will assure surprise break of shot, no reflex action of muscles to disturb sight alignment because shot is fired before any reaction to the coming recoil takes place.

  c. Third Shot: Fundamentals applied.

  d. Fourth Shot: Same.

  e. Fifth Shot: Same.

TRANSITION: The lack of communication between coach and shooter means that many errors are not detected and therefore are not corrected.

 VI. CONCLUSION:

  A. Retain Attention:

  We are the guardians of a great endeavor. We have achieved much and we have erred often. We are willing in all humility to make way for those who will follow. We seek perfection with the best that is in us. May the heritage we have sustained impel our champions into the tempest invested with unassailable fortitude.

           The Pistol Coach-Instructor-Shooter.

  B. Summarize:

   1. Coaching the Champions.

   2. Factors in systematic coaching technique.

   3. Head coach technique.

   4. Line coach technique.

   5. Suggestions for improving coaching technique.

   6. Demonstration of pistol coaching technique.

  C. Application: The results achieved will be far greater by systematically applying proper coaching techniques to control the performance of the shooter.

  D. Closing Statement: Behind every top shooter or winning team is a good coach with a commanding knowledge of how to win at shooting a pistol.

PREPARATION DATA SECTION

I.	TITLE AND NUMBER:	Techniques of coaching a pistol team (P-10)
II.	TIME ALLOTTED:	Fifty (50) minutes.
III.	PERIODS OF INSTRUCTION:	One (1)
IV.	CLASS PRESENTED TO:	Coaches and Instructor-Shooters
V.	INSTRUCTIONAL REFERENCES:	Advanced Pistol Marksmanship Manual
VI.	INSTRUCTIONAL AIDS:	Podium, podium cards, blackboard, chalk, 2 gun boxes complete with one .45 cal pistol each, firing benches, 1 pair binoculars, 3 chairs, 1 set equipment for coach (demonstration of coaching technique), GTA charts of 35mm slides & projector, 1 sound set, 1 revolver cal .38 and 5 rounds of .38 cal blank ammunition.
VII.	STUDENT EQUIPMENT AND UNIFORM:	Uniform as prescribed.
VIII.	PHYSICAL FACILITIES:	Bleachers or classroom.
IX.	PERSONNEL REQUIREMENTS:	1 principal instructor, 1 assistant instructor, (if skit is used, an additional 4 assistant instructors needed), 1 sound set operator.

PHYSICAL FITNESS I

PHYSICAL CONDITIONING (P-12)

LESSON OUTLINE

I. LESSON OBJECTIVE: To impress pistol marksmanship students with the importance of conducting a daily physical conditioning program to condition themselves, mentally and physically, to better withstand the pressures of match shooting conditions.

II. STUDENT PERFORMANCE OBJECTIVES: As a result of this instruction the student must be able to:

A. KNOW the basis for a good physical condition.

B. FORM a habit of engaging in sports and activities that are advantageous to good physical conditioning.

C. PLAN and EXECUTE a program of daily physical conditioning exercises using the pistol team daily dozen.

D. BECOME familiar with and use a daily program of static tension exercises.

III. ADVANCE ASSIGNMENT: The Advanced Pistol Marksmanship Manual, Chapter XII, "PHYSICAL CONDITIONING."

IV. INTRODUCTION: (3 Min)

A. Gain Attention:

If inhabitants of other planets were capable of tuning in on our TV Networks and viewed the commercials for one day they would in all probability classify earth as a planet of sick people. The commercials do us humans an injustice by exposing us as a race of people with headaches, backaches, earaches, sneezes, wheezes and a variety of other ailments. The source of many of our bodily aches and pains can be attributed to a general run down condition, brought on by overeating, a lack of planned nutrition, exercise and the resulting obesity. Most of us are guilty of neglecting to keep our physical condition up to par. This has been recognized by U.S. Presidential Health Councils seeking to bring to the attention of all Americans the deplorable condition of the nation's Health.

B. Orient Students:

1. Lesson Tie-In: All of us have engaged in some form of exercise from the day we took our first step and exercise became a part of growing up. Most of this exercising was done while you were still young and full of energy. The interest seemed to fade when we took on the adult burdens and responsibilities of work, marriage and raising a family until we finally succumbed to activities requiring less physical endurance. Now you are entering into a new interest, "Pistol Shooting." If you haven't already found out you will shortly find that good physical condition is an important factor in good shooting performance.

2. Motivation: "Is exercise good for you"? No one would reply negatively to this question and yet so many of us badly in need of such a program keep putting it off because of one excuse or another. Time is no excuse nor is the excuse that exercises are too exhaustive For those who become exhausted easily doing exercises such as running or the Pistol Team

Daily Dozen, we have tension type strength building exercises which can be accomplished in less than two minutes daily. They have a conditioning effect quite similar and fully as beneficial as motion type conditioners.

   3. <u>Scope</u>: During this period we will discuss the importance of good physical conditioning to the competitive pistol shooter, how to perform the pistol shooters daily dozen and the value of including static tension exercises in a daily program.

V. BODY:

  A. <u>First Student Performance Objective</u>: Students must know the basis for a good physical condition.   (5 Min)

   <u>QUESTION</u>: How does the shooter know what exercises and how many to take?

   1. Physical training should be progressive, either in repetitions performed or in resistance used. Conditioning should remain short of that sought by athletes. Violent or strenuous athletics which could result in injuries should be avoided. The competitive shooter should possess the following basic physical characteristics:

    a. An adequately developed muscular system (especially the muscles of the abdomen, arms, and legs).

    b. The ability to relax and to keep from utilizing those muscles which are not required to hold the body in the ready position or to apply pressure on the trigger.

    c. Strong breathing muscles so that breathing deeply is an easy function to permit sustaining a supply of oxygen.

    d. Quick reactions.

    e. A well developed sense of equilibrium.

    f. Precision and coordination of bodily actions. Physical conditioning must consist of coordination exercises as well as those of a general nature directed toward strengthening the muscles, toward proper breathing, and toward developing body flexibility and precision of movement.

<u>TRANSITION</u>: Now that we understand the basis for a good physical condition let's see what we can do to acquire and maintain our bodies in good physical condition.

  B. <u>Second Student Performance Objective</u>: Students should form a habit of engaging or participating in sports and activities that are advantageous to maintaining good physical condition.   (5 Min)

   <u>QUESTION</u>: Is any activity besides formal exercise beneficial?

   1. There are many types of exercises and activities that a shooter can participate in to his advantage.

    a. Walking is a very good exercise. Walk briskly if you are to get any good out of it.

    b. Golfing offers many benefits for the shooters' conditioning. As with any activity it must be done regularly if it is to aid shooting performance.

c. Bowling is a good conditioner if done regularly.

d. Swimming exercises almost all the muscles. However, it is not recommended during match shooting periods but during off season training.

e. Exercises that strengthen and build the wrist and arm muscles are recommended. Caution should be exercised in the case of weight lifting. It is not necessary to be capable of pressing 100 pounds in order to hold a pistol steady at arms length.

f. Develop the grip by using a sponge rubber ball about 3" in diameter, cut in two halves. Squeeze the ball with the shooting hand whenever you wish or wherever you may be.

g. The stronger the muscle structure is developed, the surer movement can be coordinated and positions held. Besides general conditioning practices, durable muscular tension exercises of the body, trunk, shoulder and arms make the most sense. Coordination exercises and grip exercises are in order. Physical training should take place daily for at least 15-30 minutes.

TRANSITION: If you have never followed a program of daily physical conditioning and are interested in starting one that will benefit your shooting performance, we have such a program for you.

C. Third Student Performance Objective: Students should plan and execute a program of daily physical conditioning exercises using the pistol team daily dozen. (7 Min)

QUESTION: Is it possible to condition the muscles used only in shooting?

1. The pistol team daily dozen was especially developed by a former member of the US Army Pistol Team, to condition those muscles used in pistol shooting. To the beginner who does not have a daily exercise program we recommend starting out slowly by doing only four repetitions, increasing this one each day until twelve repetitions are reached. My demonstrators will perform several of these exercises as I explain the parts of the body that will benefit from each.

a. No. 1 - warm up - 4 count exercise with starting positions of hands overhead - feet 12" apart - at count of ONE, bend the waist and knees, reaching between the legs as far back as possible, at the count of TWO, recover. Three and four are a repeat of One and Two. This exercise is a good developer of the legs, particularly the back leg muscles.

b. No. II - Body twister - 4 count exercise - starting position - bent forward at the waist, arms extended parallel with ground. At count of ONE, swing right arm so as to touch left toe, keeping shoulders and arms rigid so that twisting motion is from the waist. Another good leg developer plus the abdominal and lower back muscles benefit.

c. No. III - Push-up - 4 count exercise done from the leaning rest position - at count of ONE, bend arms out at elbows and lower the body until the chest touches the floor or ground, count of two, recover, three and four are a repeat. Keep body rigid and straight. An excellent arm and shoulder muscle developer.

d. No. IV - Back bender - 4 count exercise done from the position of hands placed in small of back, feet spread 12" apart. Count of ONE, bend at waist, keeping legs stiff and touch toes with fingertips. Count of TWO, recover, count of THREE, bend backward at waist and count of FOUR, recover. This exercise develops the legs, back and abdomen.

  e. No. IX - Side Bender - 4 count exercise done with feet spread and arms extended overhead, palms together. At count of ONE; bend at waist to the right, TWO; recover, THREE; to the left and FOUR; recover. The muscles along the sides are used in this exercise.

TRANSITION: For those, who because of their general physical makeup or who have little time for improving their bodies are unable to enter into a program such as the daily dozen, we have a program which can be done almost anywhere and one that requires little time. It is a sure means of building, conditioning and strengthening muscles.

 D. <u>Fourth Student Performance Objective</u>: Students should incorporate into their daily program, static tension exercises. (7 Min)

  1. Static tension exercises themselves cannot be regarded as an alternative to other type exercises but rather an addition. The fact that they hasten muscle conditioning and provide added strength will certainly be a factor in reaching peak physical condition. They can be done when time or conditions do not permit engaging in daily dozen type exercises. The advantages are that: (1) they can be done in a short period of time, (2) they can be done almost anywhere and at anytime, (3) they do not leave you exhausted, and most important, (4) they increase muscle size, strength and endurance.

  2. The main thing is <u>EFFORT</u>. Each exercise is performed to a count of one thousand - two thousand, and so on to six thousand with maximum pressure or stretching being applied to the full count. Part of the theory is that a muscle builds more rapidly under tension applied vigorously for a short period of time than when put to use over a long period.

  3. There are many types of static tension exercises, some that require simple exercises and others nothing more than effort. The demonstrators will perform several to give you an idea of how to do them.

   a. No. I - Hands held palms together, fingertips in line with chin, elbows raised in line with shoulders. Press hands together with as much effort as possible. A good builder and strengthener of arms, shoulders and chest muscles.

   b. No. II - Both hands at waist level, gripping one another, now with as much effort as possible squeeze hands together. This will develop grip and forearms.

   c. No. III - Arms hanging loosely, slightly bent in front of body palms upward. Suddenly with as much effort as possible contract the bicep and clench fists tightly. Good for arms and hands.

VI. CONCLUSION: (3 Min)

 A. <u>Retain Attention</u>: Good physical condition is something we all want to have but seldom do we give serious thought to acquiring and maintaining ourselves in a physically sound condition. It certainly is necessary to maintenance of good shooting performance but more important it is necessary to good health. Observe those about you who are the top shooters and you will see a person in good physical condition.

 B. <u>Summary</u>:

  1. Having a sound knowledge of the basis for a good physical condition will enable you to select activities which will improve your general health.

2. Select those sports and activities that are within your capabilities and limitations but make a habit of engaging in them regularly. Out-of-doors activities are preferable and start into them slowly until your body makes the adjustment.

3. Be serious about planning a program of daily exercise. The ladies should have no difficulty. The TV networks usually have an exercise program on daily. Provide a certain period each day to conduct your program and when that time comes drop everything. Several repetitions to start is adequate, increasing the number with time.

4. Static tension exercises can start you off very nicely and when you have rebuilt your muscle strength. Keep in mind that this is just an addition to exercises where muscle tone is derived from exercises requiring motion.

C. <u>Application</u>: If you already have a program which you follow, continue. If not then you should wait until after this match or season. Jumping into a vigorous program will have certain disadvantages to your shooting performance so go slowly at first.

D. <u>Closing Statement</u>: Exercise, together with proper diet and moderation of habits all play an important part in everyones' life. They are necessary to maintain us in good physical and mental health and one cannot acquire any without the other. An understanding of what is good and what is bad for us should be our approach to improved health. We are neither too young nor too old to start understanding more about our physical selves.

PREPARATION DATA SECTION

I.	TITLE AND NUMBER:	Physical Conditioning (P-12)
II.	TIME ALLOTTED:	Twenty-five (25) minutes
III.	PERIODS OF INSTRUCTION:	One-half (1/2)
IV.	CLASS PRESENTED TO:	Coaches and Instructor Shooters
V.	INSTRUCTOR REFERENCE:	Advanced Pistol Marksmanship Manual.
VI.	INSTRUCTIONAL AIDS:	1 Podium, podium cards, 1 sponge rubber ball, (3" dia) cut in half, 2 dumbbells (5 lbs ea) 1 wrist and forearm builder, 1 sound set, GTA charts or 35mm slides and projector.
VII.	STUDENT EQUIPMENT AND UNIFORM:	Uniform as prescribed.
VIII.	PHYSICAL FACILITIES:	Outdoor bleachers or classroom for students.
IX.	PERSONNEL REQUIREMENTS:	1 principal instructor, 1 assistant instructor, 1 sound set operator.

PHYSICAL FITNESS II

DIET AND HEALTH OF THE COMPETITIVE PISTOL SHOOTER (P-13)

LESSON OUTLINE

I. LESSON OBJECTIVE: To impress upon the pistol marksmanship students the importance of following a well planned diet that provides the body with all the nutrients essential to good health, good shooting and avoiding the consumption of food on the days of match shooting that may be detrimental.

II. STUDENT PERFORMANCE OBJECTIVES: As result of this instruction the student must be able to:

    A. EXPLAIN what part proteins, vitamins, minerals, salts, carbohydrates, and fats play in the repair and up keep of our bodies.

    B. UNDERSTAND the importance of proper diet and nutrition and plan a suitable diet for their personal use.

    C. FOLLOW a program of proper diet on the days of match shooting.

III. ADVANCE ASSIGNMENT: The advanced pistol marksmanship manual, Chapter XIII, "Diet and Health of the Competitive Pistol Shooter".

IV. INTRODUCTION: (3 Min)

    A. Gain Attention:

"It is conceded by many authorities that research on the special dietary needs of very active or hard working people and athletes is far more advanced in Russia today than in United States of America". This statement was made by US Olympic Coach Bob Hoffman in a special report that he prepared based on the scientific findings of over 100 of Russia's greatest scientists, conducted over a 25 year period. The report dealt with essential nutrients and contained a well planned diet for athletes. We can only assume that the diet Mr. Hoffman recommended was in part responsible for the success of US Athletes in the 1964 World Olympics held in Tokyo, Japan. Mr. Hoffman goes on to say that "as nutritional problems are solved, sportsmen, physicians and coaches will be able to accelerate the creation of stable physical equilibrium in athletes when they perform their exercises, thus making possible new records in every sport". This idea is especially applicable to the sport of advanced pistol shooting.

    B. Orient Students:

        1. Lesson Tie-In: Having received a period of instruction on physical conditioning we are now going to discuss another important factor to acquiring and maintaining good health, proper diet and nutrition.

        2. Motivation: Gayelord Hauser, America's famed diet and health specialist said, "I believe that you are as young as you look, feel, think, hope, believe and act. And I believe that the way you look, feel, think, hope, believe and act depends on three things:

            a. Good Food.

            b. A Strong, Eager Body.

            c. An Adventurous Spirit, in short, I believe that you are as young as your diet.

3. <u>Scope</u>: You must learn how essential diet is to good health, how to plan a diet and what to eat on the day of the match.

V. BODY:

A. <u>First Student Performance Objective</u>: Students must have a general knowledge of what part proteins, vitamins, minerals, salts, carbohydrates and fats play in the repair and upkeep of their body. (12 Min)

<u>QUESTION</u>: What is meant by good nutrition?

1. It is adequate nutrition, giving the individual cells of the body not only quantity but also the quality of nourishment they require. Second, it is balanced nutrition, supplying the body cells with vital nutrients in the proper proportion. Scientists are unanimous in agreeing that overnutrition, through excess calories stored as fat, can contribute materially to physical deterioration.

As a simplified, perhaps crude illustration, think of your body as a motor car. It is made of <u>protein</u>, inside and out. Arteries, glands, colon, connective tissue, muscles, skin, bones, hair, teeth, eyes, all contain protein and are maintained and rebuilt with <u>protein</u>. <u>Fats</u> and <u>carbohydrates</u> are your body's oil and gasoline, they are burned together to produce energy. Vitamins and minerals are its sparkplugs, essential to the utilization of food and its assimilation into the blood stream. It is a marvelously sturdy motor car, this body of yours. Marvelous in its ability to maintain and rebuild itself. Given care, consideration and respect, it will function smoothly. Neglected or abused, it will break down.

2. Mr. Hoffman's report was quite detailed but we have extracted some portions that were deemed essential information for all persons engaged in activities where speed, reaction time, coordination, and concentration are essential for peak performance.

a. On the subject of Proteins - "<u>Increased protein is vital to increased effort</u>, speed and endurance. Results of physiological studies showed that <u>foods rich</u> in <u>protein increase</u> the <u>excitability</u> of the <u>nervous system</u>, <u>enhances reflex activity</u> and <u>increases</u> the <u>speed of reaction</u>. It improves <u>ability</u> to <u>concentrate</u> with <u>considerable effort</u> for <u>limited periods</u> of time.

<u>Up to 300 grams of protein daily</u>. The use of a high protein and carbohydrate diet in this case (referring to athletes) must be followed for other reasons than in exercises of great intensity. The increase of protein consumption, up to 2.5 to 3 grams per kilo of body weight, one kilo equal to about 2.2 pounds (three times the amount listed by many authorities in this country, 300 grams of protein daily for hard training men of fairly large size) is dictated due to considerable losses of nitrogen as a result of prolonged muscular activity which under the usual protein norms may lead to a negative protein balance.

Since essentially all athletic exercises lead to various degrees of oxygen deficiency, the basic source of energy in muscular activity are carbohydrates. All of this indicates that for an athlete to <u>reach his highest potential</u>, he should go on a <u>high protein - carbohydrate</u> diet and <u>reduce fat consumption</u>.

Mr. Hoffmans recommendations for an athlete's proper diet can be modified somewhat to fit the needs of the pistol shooter. The degree of physical exertion experienced by a shooter on a day of competition is much less than that of an athlete. Therefore the shooter wouldnot require the amounts the diet suggested. However, the shooter should include in his daily diet all the nutrients essential to maintenance of a healthy body. All effects induced by overly abundant amounts of protein are not conducive to better shooting performance. An excited nervous system is a detriment.

b. On the subject of vitamins - "the increased need of vitamins" The nutrition of athletes as compared to laborers should be more abundant in vitamins as well as in proteins. According to many tests the requirements of an athletes body in vitamins rises consistently upon execution of physical exercises, apparently as a result of high metabolic intensity. If a body works harder it simply needs more fuel of every kind to keep up its accelerated pace.

TRANSITION: Athletes need more protein - more vitamins - more minerals. Since 1951, we have been recommending 300 grams of complete protein daily to hard working men and women and athletes. This shows how long we have been right. All people, we believe, need more protein, more vitamins and more minerals than many of the standards of minimum daily requirements which are being offered. We believe that in a rich country such as the US, we should not try to exist on minimums, we should supply our bodies with enough, in fact more than enough of essential nutrients.

    B. Second Student Performance Objective: The shooter must understand the importance of proper diet and nutrition and be able to plan a diet for his personal use. (5 Min)

    QUESTION: What are the most beneficial foods in a shooters' diet?

    1. Suggested items in a shooters diet. All are highly beneficial.

        a. Meats - steak, liver, beef, chicken, fish, always broiled, baked or boiled.

        b. Eggs - anyway but raw or fried.

        c. Starches - Potatoes - Boiled, baked, hash brown, augratin or scalloped. Bread - whole wheat, rye or pumpernickel.

        d. Milk - No whole milk. Skim milk is excellent as is buttermilk and yogurt. Solid cheeses are ok but no processed cheese, use powdered skimmed milk to fortify eggs, soups, milk and other dishes, it will supply considerable protein, 37% and a lot of sugar energy. It is 40% lactose or milk sugar.

        e. Fruit - Very good, especially fresh fruits and fresh juices. Never use sweetening, except honey.

        f. Vegetables - yellow and green. Double up on salads with plenty of salad oil, germ oils can be used for this purpose.

        g. Drinks - Use bottled spring water if possible, unsweetened fruit juices, skim milk, buttermilk, vegetable juices and thin soup.

        h. Desserts - Custards made with honey and skimmed milk. Fruits and berries, unsalted nuts.

        i. Salt - salt tablets during competition sometimes cause nausea. Salt your food adequately in hot weather to prevent heat prostration.

        j. Vitamins - Complete vitamin - mineral tablets, with extra vitamin C.

TRANSITION: There is a wide variety of beneficial foods readily available. All of this good food however, is not beneficial on the days in which a shooter may participate in competitive match shooting.

C. <u>Third Student Performance Objective:</u> The pistol-shooter student must follow a planned diet or rations on the days of match shooting.

<u>QUESTION:</u> Will a piece of candy give the shooter quick energy?

1. Breakfast for the athlete on the day of competition, should be a big meal for it must provide energy for the competition. Fruit, honey, two to four eggs, not oatmeal, skim milk, whole wheat or pumpernickel bread. It is not wise to attempt to shoot on an empty stomach

A word of caution - <u>Do Not</u> take <u>extra sugar</u> or <u>dextrose</u> before competing. This gives you a high rise in digestive sugar level which sets off an increased flow of body insulin which temporarily lowers the blood sugar level below normal. You can undo a week of training in five minutes.

2. A specific lunch time is not required. During the long shooting day, refresh yourself with oranges. Suck vitamin C tablets. No tobacco or alcohol, have a cookie and lemonade in limited amounts or weak tea. No tranquilizers or pep pills, unless prescribed by your doctor.

3. Eight to ten hours of sleep is recommended. Regular hours are urged. 10 PM to 6 AM if possible, will prepare him for the following day of competition. Arise early, do a few warmup exercises, shower and dress and eat breakfast early enough so that time does not become of paramount importance. Keep the pace slow and hope nothing will excite you.

4. At the conclusion of the days competition the shooter should then plan the dinner meal that will offer him all the essential nutrients. A satisfying dinner, a relaxed evening and a good night of restful sleep will provide the shooter with all the ingredients of a winning performance.

VI. CONCLUSION: (3 Min)

A. <u>Retain Attention:</u> How often have you actually sat down in a restaurant scanned the menu and planned your dinner according to the adequacy and balance of a planned diet. You order to your likes and dislikes, which is normal. You don't need to be a dietician in order to plan a nutritious dinner and enjoy it. All that is necessary is that you know what foods provide the nutrients necessary for good health.

B. <u>Summary:</u>

1. We have discussed in general terms the necessity for everyone to know what constitutes an adequate and balanced nutrition and what a diet should consist of to acquire this adequacy and balance.

      a. The elements of a proper diet.

      b. Planning a proper diet for a shooter.

      c. Diet or rations on the day of the match.

C. <u>Application:</u>

1. Too many of us ignorant of our bodies needs, live from day to day on diets that are not planned thus depriving ourselves of the benefits of inherently high physical potential. We deteriorate before our time. Planned diets are a must for everyone in the pistol shooting game, if we are to keep ourselves fit and be competitors worthy of our up keep.

D. **Closing Statement:** Good diet is necessary for building and maintaining good health just as exercise is necessary for building and maintaining muscle tone. The importance of diet to the pistol shooter is as basic as the fundamentals if one is to achieve recognition as a top competitor.

PREPARATION DATA SECTION

I.	TITLE AND NUMBER:	Diet and Health of the competitive pistol shooter (P-13)
II.	TIME ALLOTTED:	Twenty-five (25) minutes
III.	PERIODS OF INSTRUCTION:	One half (1/2)
IV.	CLASS PRESENTED TO:	Coaches and Instructor Shooters
V.	INSTRUCTOR REFERENCES:	The Advanced Pistol Marksmanship Manual
VI.	INSTRUCTIONAL AIDS:	1 podium, podium cards, 1 sound set, GTA charts or 35mm slides and a slide projector.
VII.	STUDENT EQUIPMENT AND UNIFORM:	As prescribed
VIII.	PHYSICAL FACILITIES:	Outdoor bleachers or classroom
IX.	PERSONNEL REQUIREMENTS:	1 principal instructor, 1 assistant instructor, 1 sound set operator.

PHYSICAL FITNESS III

EFFECTS OF ALCOHOL, COFFEE,
TOBACCO AND DRUGS (P-14)

LESSON OUTLINE

I. LESSON OBJECTIVE: To familiarize the student with the adverse effect alcohol, tobacco, coffee, and drugs have on shooting control.

II. STUDENT PERFORMANCE OBJECTIVES: As a result of this instruction the pistol shooter-student should be able to:

    A. EXPLAIN the detrimental effects of alcohol on shooting control.

    B. UNDERSTAND the adverse effects of coffee and tea on control of shooting.

    C. EXPLAIN the effects of tobacco use on pistol shooting and offer a method of abstaining.

    D. UNDERSTAND how the use of drugs in prescription medicines and certain home remedies are detrimental to shooting.

III. ADVANCE ASSIGNMENT: The Advanced Pistol Marksmanship Manual, Chapter XIV, Effects of Alcohol, Tobacco, Coffee and Drugs.

IV. INTRODUCTION: (3 Min)

    A. Gain Attention: The hangover, the antidote, the nerve calmer, the picker-upper, these are other names for alcohol, coffee, tobacco and drugs.

    B. Orient Students:

        1. Lesson Tie-In: You've heard the expression "built-in" error several times during your shooting experience. A few of these built-in errors are caused by intemperate habits.

        2. Motivation: The use of stimulants and depressants have varying effects on all of us. As shooters and coaches you should be aware of these effects as they pertain to shooting

        3. Scope: During this period we will discuss harmful effects of various stimulants and depressants on the body and how their use is detrimental to pistol shooting.

V. BODY:

    A. First Student Performance Objective: The pistol shooter-student should be able to explain why alcohol has a detrimental effect on shooting. (10 Min)

        QUESTION: How long does alcohol retain its effect on the body after cessation of consumption?

        1. Acts as a depressant.

        2. Dulls the senses.

3. Lessen desire to win.

4. Destroys coordination.

5. Lessens ability to concentrate.

6. Promotes carelessness.

7. Contributes to loss of judgement.

8. Dehydrates the body.

9. Involves the development of permanent damage to health.

10. Established by experience as a habit forming agent.

TRANSITION: Scientific studies have established proof of the damaging effect of alcohol on the human body. The adverse effect of coffee and tea on shooting control needs no scientific proof.

B. Second Student Performance Objective: The shooter should understand why coffee and tea have an adverse effect on shooting performance. (10 Min)

QUESTION: Does coffee increase heartbeat enough to be noticeable while shooting?

1. Acts as a stimulant.

2. Stimulates increased heart beat.

3. Adversely affects nervous system as regards control of shooting.

4. Similar effects of tea.

 a. Caffeine.

 b. Tannic acid.

5. Cola drinks contain caffeine which is highly stimulating.

TRANSITION: Alcohol and coffee are relatively temporary in the duration of their adverse effect on shooting but the user of tobacco is under the effect of his hard-to-break habit practically 24 hours a day for an indefinite period.

C. Third Student Performance Objective: The student must appreciate the bad effect tobacco has on the competitive pistol shooter. (20 Min)

QUESTION: How long after the last cigarette does the shooter continue to feel the effects?

1. Nervous system disturbances are continuous and damaging to shooting results.

2. Nicotine is a deadly poison.

3. Lung damage is inevitable after protracted use of tobacco.

4. Circulatory system damage is great enough to cause the heart to overwork.

5. Increased heart beat is noticeable during shooting.

6. Loss of appetite and weight.

7. Expense.

8. Danger of other serious illness or death.

9. Habit forming properties established beyond doubt.

10. You can quit smoking.

11. Commentary by Dr. Crane in the newspaper "Arizona Republic," 1961.

12. The Surgeon General's Report on smoking, 11 Jan 1964.

TRANSITION: Many shooters that are aware of their ill effects and as a result avoid tobacco, alcohol and coffee innocently fall prey to the devastating effect of certain drugs found in commonly used medicines and pills.

D. Fourth Student Performance Objective: The pistol shooter should know how certain drugs found in medicines and pills adversely affect shooting. (5 Min)

QUESTION: If a shooter has a bad cold, should he accept treatment before shooting?

1. Depressants slow down the bodily processes in the following manner.

   a. Slow reflexes are detrimental to fine coordination.

   b. Alertness is affected by the slowed down mental state.

   c. Lessens the desire to win.

   d. Loss of ability to concentrate intensely for the accustomed length of time.

   e. Promotes carelessness in adhering strictly to techniques of control.

   f. Usually found in sleeping pills and other barbiturate preparations.

2. Stimulants abnormally accelerate the bodily processes and give a false sense of well being.

   a. A feeling of nervousness and a slight trembling are the usual symptoms.

   b. Increased anxiety is experienced concerning results of the match.

   c. Involuntary movement of the extremities (excessive movement of hands).

   d. Increases heart beat to a noticeable level.

   e. Some are habit forming as are alcohol and tobacco.

   f. Usually found in keep awake pills, benzedrine inhalers, appetite curbers, etc.

3. Drugs in daily use.

   a. Sedatives and depressants.

   b. Analgesics (pain relief).

   c. Antihistamines (relief of colds and associated illness)

   d. APC tablets (same as above).

   e. Decongestants (relief of nasal and sinus congestion, etc.)

VI. CONCLUSION: (2 Min)

   A. <u>Retain Attention</u>: How many students in this class have stopped smoking since the first of the year?

   B. <u>Summarize</u>:

   1. Alcohol.

   2. Coffee and Tea.

   3. Tobacco.

   4. The use of drugs.

   C. <u>Application</u>: Try giving up alcohol and tobacco, stop prescribing remedies for your aches and pains. Desist completely or try to reduce consumption of coffee drasticly for several months. Decaffeinated coffee will help. See if you don't feel better and shoot better.

   D. <u>Closing Statement</u>: Remember that to be a fine pistol shot you must train diligently. A good athlete undergoes rigorous training to develop his ability--and indulge in nothing that will prevent the attainment of top honors.

PREPARATION DATA SECTION

I.	TITLE AND NUMBER:	The effects of Alcohol, Coffee, Tobacco and Drugs (P-14)
II.	TIME ALLOTTED:	Fifty (50) minutes
III.	PERIOD OF INSTRUCTION:	One
IV.	CLASS PRESENTED TO:	Coaches and Instructor Shooters
V.	INSTRUCTOR REFERENCES:	The Advanced Pistol Marksmanship Manual and "Alcohol, Its Effect on Man," by Haven Emerson MD; "The Cigarette and You", by Donald W. Hewitt, MD; "Drugs in Daily Use" by Walter Modell, MD.
VI.	INSTRUCTIONAL AIDS:	Podium, podium cards blackboard, chalk, pointer, GTA charts or 35mm slides and slide projector.
VII.	STUDENT EQUIPMENT AND UNIFORM:	Uniform as prescribed.

VIII. PHYSICAL FACILITIES: Outdoor bleachers or classroom for students.

IX. PERSONNEL REQUIREMENT: One (1) principal instructor, 1 asst instructor 1 sound set operator.

REVIEW OF COURSE OF INSTRUCTION AND WEAK POINTS (I-8)

LESSON OUTLINE

I. LESSON OBJECTIVE: To reemphasize the important points covered in the entire Advanced Pistol Marksmanship Course of Instruction.

II. STUDENT PERFORMANCE OBJECTIVES: As a result of this instruction, the advanced pistol shooter or coach should be able to:

    A. FUNDAMENTALS I (Attain a minimum arc of movement)

    B. FUNDAMENTALS II (Control sight alignment)

    C. FUNDAMENTALS III (Understand trigger control factors and methods)

    D. TECHNIQUE OF FIRE I (Establish a system)

    E. TECHNIQUE OF FIRE II (Learn technique of Slow Fire)

    F. TECHNIQUE OF FIRE III (Learn technique of Sustain Fire)

    G. UNDERSTAND the value of mental discipline to the advanced pistol shooter.

    H. COACHING I (Explain the attributes, duties and responsibilities of a pistol team coach)

    I. COACHING II (Know the technique of coaching a pistol team)

    J. COACHING III (Evaluate the team shooter)

    K. PHYSICAL FITNESS I (Know how to physically condition the competitive pistol shooter)

    L. PHYSICAL FITNESS II (Control the diet and health of the competitive pistol shooter)

    M. PHYSICAL FITNESS III (Understand the effects of alcohol, coffee, tobacco and drugs on the pistol shooter)

    N. COMPETITIVE REGULATIONS I & II (Understand NRA Pistol Match Rules, Range Procedure, and Range Safety)

    O. COMPETITIVE REGULATIONS III & IV (Explain the National Trophy Pistol Match Rules and Requirements for earning a Distinguished Pistol Shot Badge.)

    P. ADMINISTRATION I (Disclose the team organization and administration of a pistol team)

    Q. ADMINISTRATION II (Procure, maintain and safeguard match weapon, equipment and ammunition for a pistol team)

III. ADVANCE ASSIGNMENT: None

IV. INTRODUCTION: (2 Min)

    A. Gain Attention:

        1. Lesson Tie-In: As we approach the end of the course, we will refresh your memory on the most important points of advanced pistol marksmanship.

        2. Motivations: You must have a thorough knowledge of the important factors of advanced pistol shooting in order to shoot with or properly train a winning pistol team.

        3. Scope: During this period we will review the complete course of advanced pistol marksmanship instruction.

V. BODY:

    A. First Student Performance Objective: Fundamentals I (Attaining a Minimum Arc of Movement) (4 Min)

        QUESTION: How can the shooter know when he has a good stance?

        1. Stance

        2. Position

        3. Grip

        4. Breath control

    B. Second Student Performance Objective: Fundamentals II (Control of Sight Alignment) (4 Min)

        QUESTION: On what point does the eye focus during aiming?

        1. Relationship of sights.

        2. Point of focus

        3. Concentration

        4. Characteristics of the human eye

    C. Third Student Performance Objective: Fundamentals III (Understand Trigger Control Factors and Methods) (4 Min)

        QUESTION: What is positive trigger pressure?

        1. Nerve processes

        2. Factors providing for the correct control of the trigger

        3. Application of positive trigger pressure

    D. Fourth Student Performance Objective: Technique of Fire I (Establishing a System) (4 Min)

QUESTION: What is an effective shooting system based on?

1. Preparation for shooting

2. Plan shot sequence

3. Relaxation before shot or string

4. Deliver shot or string of shots on target

5. Make an analysis of each shot or string

6. Make a positive correction if necessary

 E. <u>Fifth Student Performance Objective:</u> Technique of Fire II (Learn Technique of Slow Fire) (4 Min)

QUESTION: Does recoil of the pistol have any effect on accuracy?

1. Employment of the fundamentals

2. Techniques in slow fire control

3. Common deficiencies in control

4. Wind shooting and adverse conditions

5. Training methods

 F. <u>Sixth Student Performance Objective:</u> Technique of Fire III (Learn Technique of Sustained Fire) (4 Min)

QUESTION: What is the technique of control in rapid fire when the wind is blowing?

1. Employment of the fundamentals

2. Techniques of sustained fire

3. Common deficiencies in control

4. Training methods

5. Wind shooting and adverse conditions

 G. <u>Seventh Student Performance Objective:</u> Understand the value of mental discipline to the shooter. (4 Min)

QUESTION: What is mental discipline?

1. Essential to advance pistol marksmanship

2. Developing mental discipline and confidence

3. Why can't you be a winner? or the danger or negative thinking.

4. Match pressure

5. Reducing tension and attaining relaxation

   H. <u>Eighth Student Performance Objective</u>: Coaching I (Explain the Attributes, Responsibilities and Duties of a Pistol Team Coach)

   <u>QUESTION</u>: How much improvement should a pistol shooter show to safeguard his position?

   1. Personal attributes of a coach

   2. Head coach responsibilities

   3. Line coaches duties

   I. <u>Ninth Student Performance Objective</u>: Coaching II (Technique of Coaching a Pistol Team) (4 Min)

   <u>QUESTION</u>: How much free practice time should a pistol shooter have to work on weaknesses?

   1. Coaching the champions

   2. Factors in systematic coaching technique

   3. Head coach technique

   4. Line coach technique

   5. Suggestions for improving coaching technique

   6. Demonstration of pistol coaching technique

   J. <u>Tenth Student Performance Objective</u>: Coaching III (Evaluate the Pistol Team Shooter) (4 Min)

   <u>QUESTION</u>: How often should you change the composition of your top team?

   1. Individual information and evaluation sheet

      a. Line coach evaluation

      b. Progress graph

      c. Aggregate record

   2. Examination

   3. Score book

   4. Attributes of a team shooter

   K. <u>Eleventh Student Performance Objective</u>: Physical Fitness I (Know how to Physically Condition the Competitive Pistol Shooter) (4 Min)

QUESTION: Is it best to exercise in the morning or afternoon after firing?

    1. Basis for good physical condition

    2. Types of exercises

    3. Pistol team daily dozen exercises

    4. Static tension exercises

L. Twelfth Student Performance Objective: Physical Fitness II (Control the Diet and Health of the Competitive Pistol Shooter) (4 Min)

QUESTION: What makes up a good lunch for a shooter during matches?

    1. General information

    2. The importance of proper diet and nutrition

    3. Diet and care of health - Dr. Granjean of Switzerland

M. Thirteenth Student Performance Objective: Physical Fitness III (Understand the Effects of Alcohol, Coffee, Tobacco and Drugs on the Pistol Shooter) (4 Min)

QUESTION: How long does the effect of a cigarette last?

    1. Alcohol

    2. Coffee and tea

    3. Tobacco

    4. Drugs

N. Fourteenth Student Performance Objective: Competitive Regulation I and II (Understand NRA Pistol Match Rules, Range Procedure and Range Safety) (4 Min)

QUESTION: Can the shooter plug his own shots when scoring?

    1. Types of competition

    2. Competition regulations

    3. Challenges and protests

    4. Match weapons, equipment and ammunition

    5. Scoring

    6. Range control commands and safety

    7. Competitors duties and responsibilities

    8. Team officers duties and position

9. Eligibility of competitors

O. <u>Fifteenth Student Performance Objective</u>: Competitive Regulations III and IV (Explain National Trophy Pistol Match Rules and Requirements for Earning a Distinguished Pistol Shot Badge) (4 Min)

P. <u>Sixteenth Student Performance Objective</u>: Administration I and II (Disclose the Team Organization and Administration, Procurement, Maintenance and Security of Weapons, Equipment and Ammunition)

<u>QUESTION</u>: Is there any authority for company level Army units to obtain weapons for competition?

1. Team Organization and administration

2. Procurement sources

3. Causes of weapons malfunctions and care and maintenance of weapons

4. The Marksmanship Unit Shop is part of the team effort

Q. <u>Seventeenth Student Performance Objective</u>: Review Weak Points in Shooting Performance. (10 Min)

1. Improper grip

2. Improper position and stance

3. Early first shot

4. Lack of rhythm

5. Lack of follow thru and natural recovery

6. Holding too long

7. Looking at target

8. Anticipation-heeling

9. Impatience-jerking

10. Lack of complete shot analysis

11. Do not shoot and practice alone

12. Unresolved shooting problems

13. Don't knock it until you have tried it

VI. CONCLUSION: (1 Min)

A. <u>Retain Attention</u>: Any misunderstanding or lack of knowledge of pistol marksmanship principles will be a handicap on future performance.

B. **Summarize**: Here, once more, are the major points by which we hope to develop in this nation a large nucleus of advanced pistol shooters and coaches.

C. **Closing Statement**: Knowledge of the advanced pistol marksmanship fundamentals gives the shooter a good foundation on which to build good habits and good scores. Improvement beyond the basic accomplishments is initiated by acquainting the shooter with the highly satisfactory results to be obtained through developing the necessary mental discipline. Successful, consistent control in applying the fundamentals is the secret of championship shooting.

PREPARATION DATA SECTION

I.	TITLE AND NUMBER:	Review of course of instruction and weak points (I-10)
II.	TIME ALLOTTED:	Seventy-five (75) minutes
III.	PERIODS OF INSTRUCTION:	One and one-half (1 1/2)
IV.	CLASS PRESENTED TO:	Coaches and instructor shooters
V.	INSTRUCTOR REFERENCES:	Advanced Pistol Marksmanship Manual
VI.	INSTRUCTIONAL AIDS:	1 Podium, podium chart, 1 blackboard, .45 cal pistol, chalk, GTA chart or 35mm slides and projector.
VII.	STUDENT EQUIPMENT AND UNIFORM:	Uniform as prescribed.
VIII.	PERSONNEL REQUIREMENTS:	1 Principal instructor, 1 asst instructor, 1 Sound set operator
IX.	PHYSICAL FACILITIES:	Classroom or outdoor bleachers.

WRITTEN EXAMINATION AND CRITIQUE WITH PANEL DISCUSSION (I-9)

LESSON OUTLINE

I. LESSON OBJECTIVE: To test students in advanced pistol marksmanship, critique and discuss the important points.

II. STUDENT PERFORMANCE OBJECTIVES: As a result of this instruction the student should be able to:

    A. CONDUCT a written examination.

    B. MODERATE a critique.

    C. LEAD a panel discussion.

III. ADVANCE ASSIGNMENT: The Advanced Pistol Marksmanship Manual, Vol I.

IV. INTRODUCTION: (3 Min)

    A. <u>Gain Attention</u>: Pass out one test booklet per student.

    B. <u>Orient Students</u>:

        1. <u>Lesson Tie-In</u>: This test is designed to help the student find out where he is weak in the pistol marksmanship course of instruction. It also permits the instructor to know how well his instruction is being learned by the student. If any areas of the instructor course appear to need further stress, those portions will be argumented by additional instruction.

        2. <u>Motivation</u>: Any misunderstanding or lack of knowledge of pistol marksmanship principles will be a handicap on performance.

        3. <u>Scope</u>: During this period you will take the written test with twenty-five (25) minute time limit. After the written test, twenty (20) minutes will be allowed for a reading of the correct answers and you may question all of the instructors you had during this course of instruction. They will be available for questions or clarification in the form of a panel. Questions will be directed to the individual member of the panel who conducted the instruction. He has a commanding knowledge of that particular subject.

V. BODY:

    A. <u>First Student Performance Objective</u>: Conducting a written examination: An example. (25 Min)

    B. <u>Second Student Performance Objective</u>: Moderating a critique of written test. The correct answer will be given for each question or situation included in test. Discussion of points brought up by students will be encouraged. (20 Min)

    C. <u>Third Student Performance Objective</u>: Leading a panel discussion of points covered by the written test, the Advanced Pistol Marksmanship Course and those questions concerning the course of instruction that are posed by the students. (Conducted in conjunction with the critique)

VI. CONCLUSION: (2 Min)

    A. <u>Retain Attention:</u> The shooter who fires a possible ten shot string of rapid fire and doesn't know how it was accomplished will have difficulty duplicating the performance deliberately. The coach finds out how much you know by testing.

    B. <u>Summarize:</u>

        1. Conducting a written examination.

        2. Moderating a critique of the examination.

        3. Leading a panel discussion.

    C. <u>Closing Statement:</u> A valid test will identify for the instructor and the student shooters, the points that are not throughly understood. Subsequent instruction will stress the weak points.

PREPARATION DATA SECTION

I.	TITLE AND NUMBER:	Written examination and critique with panel discussion (I-11)
II.	TIME ALLOTTED:	Fifty (50) minutes
III.	PERIODS OF INSTRUCTION:	One (1)
IV.	CLASS PRESENTED TO:	Coaches and instructor-shooters
V.	INSTRUCTOR REFERENCE:	Advanced Pistol Marksmanship Manual
VI.	INSTRUCTIONAL AIDS:	Podium, podium cards, blackboard, chalk and pointer, one (1) examination booklet. (To be distributed to each student), GTA chart.
VII.	STUDENT EQUIPMENT AND UNIFORM:	Uniform as prescribed
VIII.	PHYSICAL FACILITIES:	Classroom, table & chair, and pencils for students.
IX.	PERSONNEL REQUIREMENTS:	1 Principal Instructor, 4 Assistant Instructors 1 Sound Set Operator.

ADMINISTRATION I AND II

TEAM ORGANIZATION AND ADMINISTRATION (I-12)

LESSON OUTLINE

I. LESSON OBJECTIVE: To familiarize the student with the Marksmanship Unit organization, administrative procedures; procurement, the necessary maintenance and security of match weapons, equipment and ammunition.

II. STUDENT PERFORMANCE OBJECTIVES: As a result of this instruction the student should be able to:

A. UNDERSTAND team Organization and Administration of the Pistol Division.

B. EXPLAIN how to use and to what extent normal U.S. Army Procurement Sources will supply the needs of a Pistol Team.

C. EXPLAIN causes of Weapons Malfunctions and Care and Maintnenance of Weapons.

D. EXPLAIN how the Marksmanship Unit Shop is Part of the Team Effort.

III. ADVANCE ASSIGNMENT: The Advanced Pistol Marksmanship Training Course Instructor's Manual, Volume II.

IV. INTRODUCTION: (2 Min)

A. <u>Gain Attention</u>: The head coach or officer-in-charge of a pistol team must know all the details of administration, supply and maintenance of a pistol team.

B. <u>Orient Students</u>:

1. <u>Lesson Tie-In</u>: Some students have become proficient in the field of marksmanship. A marksmanship unit of this type will offer certain advantages to the shooter regarding the support of his efforts.

2. <u>Motivation</u>: Knowledge of the fundamentals and advanced techniques to improve your proficiency as marksman may be wasted to some degree because of necessary involvement with administrative details.

3. <u>Scope</u>: During this period we will discuss team administration, procurement sources, causes of malfunctions, maintenance and security of weapons, equipment and ammunition.

V. BODY:

A. <u>First Student Performance Objective</u>: The pistol shooter-student must understand the team organization and administration of a pistol team. (10 Min)

<u>QUESTION</u>: How many line coaches are needed by a pistol team?

1. Organization of a pistol team.

2. Administrative responsibilities:

  a. Function of various personnel in pistol division.

 3. Administrative requirements to be met in operating a pistol team.

B. <u>Second Student Performance Objective</u>: The pistol shooter-student must know how to use normal U.S. Army procurement sources to supply a pistol team. (15 Min)

 <u>QUESTION</u>: Can a company level U.S. ARMY organization procure .22 cal pistols for competition?

 1. Table of allowances 60-18, dated 26 Jun 59, w/changes.

 2. Appropriated Funds and Non-Appropriated Funds.

 3. Expendable items from self-service store.

 4. Ear plugs and prescription glasses thru medical channels.

 5. Ammunition.

 6. Match Grade Weapons.

  a. .22 Cal.

  b. Center Fire Pistol 38 Cal.

  c. .45 Cal Service Pistol.

  d. .45 Cal Commercial Pistol (Wad-cutter).

  e. Miscellaneous Accessories.

C. <u>Third Student Performance Objective</u>: The pistol shooter-student should know the common causes of malfunctions, care and maintenance of weapons. (10 Min)

 <u>QUESTION</u>: What is the most frequent cause of pistol malfunction?

 1. Causes of malfunctions.

 2. Care and cleaning.

 3. Lubrication.

 4. Cleaning materials.

 5. Transporting weapons.

 6. Security.

D. <u>Fourth Student Performance Objective</u>: The pistol shooter-student should know why the Marksmanship Unit Shop is part of the team effort. (10 Min)

 <u>QUESTION</u>: How does the shooter determine the accuracy of his pistol?

 1. Accurizing Issue Weapons.

2. Bench Testing for accuracy.

3. Major repairs.

4. Preventive maintenance.

5. Special adjustments at the desire of the individual shooter.

6. Testing and selection of ammunition.

7. Shot group areas of dispersion.

VI. CONCLUSION:

A. <u>Retain Attention</u>: The pistol shooter's job is to shoot to win pistol matches. The job of the coach and the officer-in-charge is to make sure they think of nothing else.

B. <u>Summarize</u>:

1. Understand team organization and Administration of a pistol team.

2. Knowing how to use normal U.S. Army procurement sources to supply a pistol team.

3. Knowing the common causes of malfunctions and care and maintenance of weapons.

4. Knowing why the Marksmanship Unit shop is part of the team effort.

C. <u>Application</u>: The proper administration and supply of a pistol team will contribute immeasurably to the success of the team effort.

D. <u>Closing Statement</u>: The efficiency with which the administration and supply of a unit is carried out will free the shooter's mind so that he will develop his marksmanship talents to the greatest degree.

PREPARATION DATA SECTION

I.	TITLE AND NUMBER:	Team Organization and Administration, Procurement Maintenance and Security of Match Weapons, Equipment and Ammunition (I-15)
II.	TIME ALLOTTED:	Fifty (50) minutes
III.	PERIODS OF INSTRUCTION:	One
IV.	CLASS PRESENTED TO:	Coaches and Instructor Shooters
V.	INSTRUCTOR REFERENCES:	Advanced Pistol Marksmanship Training Course Instructor's Manual - Volume II
VI.	INSTRUCTION AIDS:	Blackboard, chalk, eraser, podium cards, notes, lesson plan, watch, pointer, GTA charts, or 35mm slides and projector, sound set, TA 60-18, TA 23-100 w/changes, gun box, scope, .45 Auto, 38 super, H. Standard S&W, or .22 Luger, shooting glasses, carbide light, lens tissue, carbide, oil, cleaning patches, cleaning rods, shaker w/resin, cleaning brushes, wrenches (bushing, trigger shoe and muzzle brake), etc.
VII.	STUDENT EQUIPMENT & UNIFORM:	Uniform as prescribed. Equipment, pencil & note book.
VIII.	PHYSICAL FACILITIES:	Classroom or bleachers
IX.	PERSONNEL REQUIREMENTS:	1 Instructor, 1 Asst Instructor, 1 Sound Set Operator

COMPETITIVE REGULATIONS I

NRA PISTOL MATCH RULES (I-14)

LESSON OUTLINE

I. LESSON OBJECTIVE: To familiarize pistol instructor-shooters and/or pistol coaches with NRA pistol match rules.

II. STUDENT PERFORMANCE OBJECTIVES: As a result of this instruction the pistol shooter-student should be able to:

A. EXPLAIN the many types of competition for pistol.

B. EXPLAIN NRA pistol competition regulations.

C. MAKE challenges and protest when necessary and know the difference between the two.

D. USE the proper match weapons, equipment and ammunition in a pistol match.

E. UNDERSTAND scoring of pistol match targets.

F. UNDERSTAND the competitor's duties and responsibilities.

G. HAVE knowledge of the team officers duties and position during a pistol team match.

H. EXPLAIN the eligibility of competitors in a pistol match.

III. ADVANCE ASSIGNMENT: The Advanced Pistol Marksmanship Training Course Instructors Manual - Volume II, Chapter XIV, "NRA Pistol Match Rules".

IV. INTRODUCTION: (3 Min)

    A. Gain Attention: The NRA referee's decision is final except in the National Matches.

    B. Orient Students:

        1. Lesson Tie-In: In order to compete in a Pistol Match it is essential that all the competitors have a complete knowledge of the rules.

        2. Motivation: Knowledge of NRA match rules will help preclude any injustice being done to you or your team; also it will help prevent mistakes for which the team could be disqualified.

        3. Scope: During this period we will discuss the different types of competition, the competition regulations with which the shooter frequently comes into contact, challenges and protests, types of weapons. how to score, competitors duties, team officers duties and eligibility of competitors.

V. BODY:

    A. First Student Performance Objective: The pistol shooter-student should know the many types of competition. (Rule I) (5 Min)

        QUESTION: Can a pistol match sponsor turn down an entry in an open pistol match?

1. (1.2) Open Match.

2. (1.2.1) National Trophy Matches (See requirements for earning a Distinguished Pistol Shot Badge - Chapter XIX - Volume II.

3. (1.2.2) National Matches.

4. (1.3) Restricted Match.

5. (1.4) Classified Match.

6. (1.5) Invitational Match.

7. (1.6) NRA Competition.

8. (1.7) League Competition.

9. (1.8) Squadded Individual Match.

10. (1.11) Squadded Team Match.

11. (1.13) Aggregate Matches.

12. (1.13.1) Tournament.

B. <u>Second Student Performance Objective:</u> The pistol shooter-student should know NRA pistol competition regulations. (Rule 9)   (10 Min)

QUESTION: How is a .45 cal pistol properly cleared?

1. (9.1) Pistols unloaded.

2. (9.2) Actions open.

3. (9.3) Pistols loaded.

4. (9.4) Cease firing.

5. (9.5) Not ready.

6. (9.6) Changing pistols.

7. (9.9) Defective cartridge.

8. (9.10) Disabled pistol.

9. (9.11) Malfunction.

10. (9.14) Weighing triggers.

11. (9.16) Competitors position.

12. (9.17) Individual coaching.

13. (9.22) Interruption of fire.

14. (9.23) Failure to function in slow fire.

15. (9.24) Failure to function in timed and rapid fire.

16. (9.25) Interference with target.

17. (9.29) Score and classification falsification.

18. (9.32) Disorderly conduct.

19. (9.33) Refusal to obey.

20. (9.34) Evasion of rules.

21. (9.35) Disqualification.

22. (9.36) Suspension.

    C. <u>Third Student Performance Objective</u>: The pistol shooter-student should know how to make challenges and protests. (Rule 16)   (5 Min)

    <u>QUESTION</u>: What is the difference between a challenge and a protest?

1. (16.1) Challenges.

2. (16.2) Protests.

3. (16.3) How to protest.

4. (16.4) Team Matches.

    D. <u>Fourth Student Performance Objective</u>: The pistol shooter-student must know the permitted match weapons, equipment and ammo. (Rule 3)   (5 Min)

    <u>QUESTION</u>: What is the allowed trigger weight of a .45 cal service pistol in a center fire match.

1. (3.1) Service Pistol

2. (3.2) Pistol cal .45 (Wad Cutter)

3. (3.3) Service Revolver cal .45

4. (3.4) Any Center Fire Pistol or Revolver

5. (3.5) .22 Caliber Pistol or Revolver

6. (3.8) Spotting Scopes

7. (3.9) Shooting Kits

8. (3.9.1) Deflecting screens

9. (3.17) Ammunition

E. **Fifth Student Performance Objective:** The pistol shooter student should understand all aspects of scoring. (Rule 14) (10 Min)

    QUESTION: Does the competitor have to sign a protested score card?

1. (14.1) When to Score

2. (14.2) Where to Score

3. (14.3) How to Score

4. (14.4) Misses

5. (14.5) Early or Late Shots

6. (14.6) All Shots Count

7. (14.7) Hits on Wrong Target

8. (14.8) Ricochets

9. (14.9) Visible Hits and Close Groups

10. (14.10) Excessive Hits

11. (14.14) Scorer's Duties

12. (14.15) Score Cards

13. (14.16) Erasures on Score Cards

14. (18.14) Signatures

15. (10.5) & (18.3.1) Competitors Will Score

16. Decisions of Ties (Rules 15)

    a. (15.3) Single stage ties

    b. (15.4) Multiple stage ties

    c. (15.5) Aggregate match ties

    d. (15.7) Team Match ties

    e. (15.10) Unbreakable ties

    f. (9.12) Perfect Score Ties

F. **Sixth Student Performance Objective:** The pistol shooter-student must understand the Competitors duties and responsibilities. (Rule 18) (10 Min)

    QUESTION: If a notice of program change is posted and the competitor doesn't know of it, is he held responsible?

1. (18.1) Discipline

2. (18.2) Knowledge of program

3. (18.3) Eligibility

4. (18.3.1) Competitors will score

5. (18.4) Classification

6. (18.5) Individual Entries

7. (18.6) Squadding Tickets

8. (18.7) Reporting at Firing Point

9. (18.8) Timing

10. (18.9) Loading

11. (18.10) Cease Firing

12. (18.11) Checking Score Card and Signing

13. (18.12) Clearing the Firing Point

14. (18.13) Checking Bulletin Board

15. (18.14) Score Cards Must be Signed

16. (18.15) Responsibility

G. <u>Seventh Student Performance Objective:</u> The pistol shooter-student should be familiar with the team officer's duties and positions. (Rule 12) (5 Min)

QUESTION: If two members of a team are firing in a team match, can both have a coach?

1. (12.1) Team Captain

2. (12.2) Team Coach

3. (12.3) Team Entries

4. (12.4) Team Captain and Coach Position

5. (12.5) Team Coach in Team Matches

H. <u>Eighth Student Performance Objective:</u> The pistol shooter-student should know the eligibility of Competitors. (Rule 2) (5 Min)

QUESTION: May a competitor who is not an NRA member, fire in the national matches?

1. (2.1) Members of NRA

2. (2.2) Civilian

3. (2.3) Junior

4. (2.4) Police

5. (2.5) National Guard

6. (2.6) Regular Service

7. (2.7) Reserve

8. (2.8) College

9. (2.9) School

10. (2.10) Team Representative

11. (2.13) Regular, N.G. or Reserve Teams

12. (2.20) Residence

VI. CONCLUSION: (2 Min)

    A. Retain Attention: If a pistol competitor is one second late arriving at the firing line he can lose as many as three hundred points from his score.

    B. Summarize:

1. Types of Competition

2. Competition Regulations

3. Challenges and Protests

4. Match Weapons, Equipment and Ammunition

5. Scoring

6. Competitor's Duties and Responsibilities

7. Team Officer's Duties and Positions

8. Eligibility of Competitors

    C. Application: NRA Pistol Match Rules will be used in all the different shooting phases of this course, in all NRA registered matches and most of the service sponsored pistol matches.

    D. Closing Statement: A competitor who has full knowledge of the pistol match rules has relieved his mind of anxiety during and after the shooting phases as to whether any of these numerous essentials have been overlooked. This will allow him to concentrate on controlling his shooting to the fullest extent of his ability.

PREPARATION DATA SECTION

I.	TITLE AND NUMBER:	NRA Pistol Match Rules (I-16)
II.	TIME ALLOTTED:	100 minutes
III.	PERIODS OF INSTRUCTION:	TWO Periods
IV.	INSTRUCTOR REFERENCES:	Coaches and Instructor Shooters

V.	INSTRUCTOR REFERENCES:	The Advanced Pistol Marksmanship Instructors' Manual - Volume II and NRA Pistol Rule Book (Current Issue)
VI.	INSTRUCTIONAL AIDS:	1 Podium, Podium Cards, 1 Sound Set, 35mm Slides, 1 - 35mm Projector or GTA Charts, 1 NMC Score Card per student, Blackboard, Pointer, Lesson Plan and Hand Notes. One - .45 Cal Pistol, Five .45 Cal Blank Cartridges, 1 NRA Pistol Rule Book per student, if available
VII.	STUDENT EQUIPMENT AND UNIFORM:	Uniforms as prescribed.
VIII.	PHYSICAL FACILITIES:	1 Classroom with Blackboard and Chalk
IX.	PERSONNEL REQUIREMENTS:	1 Prin Instr, 2 Asst Instructors (Sound & Proj Oper).

SECTION THREE
ADMINISTRATION

CHAPTER XII

PISTOL TEAM ORGANIZATION AND ADMINISTRATION

A. The organization and administration of a pistol team requires extensive planning if highly qualified personnel and optimum results are to be obtained. As a minimum requirement the pistol division of a marksmanship unit should include the following:

1. <u>ORGANIZATION.</u>

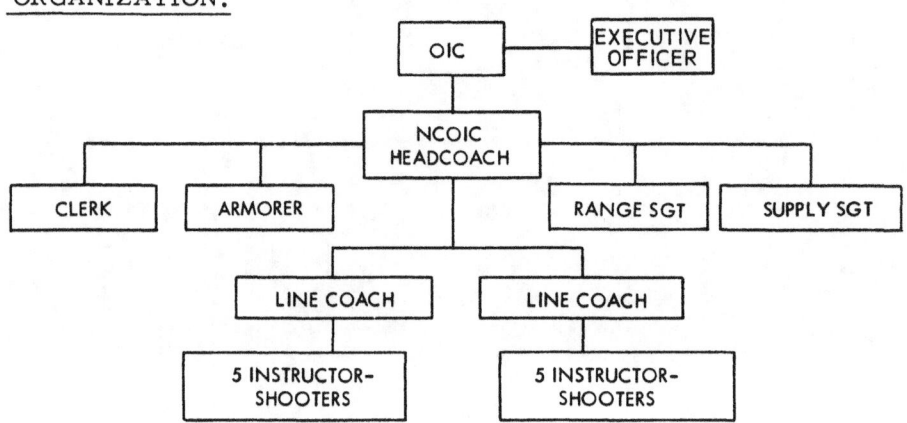

2. <u>ADMINISTRATIVE RESPONSIBILITIES.</u>

    a. OFFICER-IN-CHARGE

        (1) Plans, directs and supervises the performance of the Pistol Division.

        (2) Recommends matches for participation in and by the Pistol Division.

        (3) Makes recommendations to the Commanding officer concerning training policy and keeps him informed of the activities of the Pistol Division.

    b. EXECUTIVE OFFICER, PISTOL DIVISION:

        (1) Plan and accomplish the necessary procedures for entering all competition and furnish reports of actions and competitive results.

        (2) Compile statistics and maintain available file system.

        (3) Document policy and training doctrine by publication of directives, schedules, instructional manuals.

        (4) Supervises organization of instructor courses and conducts rehearsals of all instructors prior to training course.

        (5) Control by direct observation, the maintenance of prime weaponry in the hands of the shooters and maintain direct supervision of weapons security, turn-in and issue.

        (6) Conducts the administration of the division by preparation, forwarding and filing of official correspondence.

        (7) Participates in registered competition and practical training activities.

        (8) Supervises the accomplishment of administrative requirements.

c. HEAD COACH, PISTOL DIVISION:

(1) Assists in the planning of training.

(2) Schedules training.

(3) Directs the training by personal supervision.

(4) Initiates shooting equipment checks that serve to pinpoint faulty mechanical functioning, cleanliness, proper lubrication, security and inherent accuracy. Accountability is maintained through periodic check of equipment serial numbers.

(5) Constantly maintain a current evaluation of each shooter regularly assigned to the USAMTU Pistol Division. In addition, an evaluation and an estimate of potential should be exercised concerning outstanding shooters in major command level competition.

(6) Conduct a continuous analysis of the rate of team progress, note individual morale and attitudes and the degree of team effort exercised, all of which should be current and decisive.

(7) Propagate doctrine and performance standards by constant review of training, manuals, training materials and methods so as to reflect new ideas and methods proven to be sound and reliable and weed out unsound techniques.

(8) Improve the team potential by conducting periodic organized, group instruction and tests. Improve the individual potential by personal and private interview and conducting individual coaching sessions.

(9) Supervises the team preparation for match participation.

(10) Supervises the coaching technique of individual line coaches.

(11) Assists in preparation of instructor courses.

(12) Assists in rehearsal of instructors.

(13) Participates in all registered competition.

(14) Exercises a profound influence on the morale, attitudes and promote an enthusiatic desire of the shooter to win by exhibiting individual consideration, stimulation of confidence and creation of an atmosphere of inevitable success.

d. NON-COMMISSIONED OFFICER-IN-CHARGE (OPERATIONS SGT):

(1) In charge of team administration.

(2) Accountable to OIC of the duty status of each individual in Pistol Division.

(3) Responsible for team preparation for match competition and daily practice.

(4) Supervises and participates in organized group instruction. Assists in rehearsal of instructors.

(5) Personally supervises security, maintenance and accountability of all weapons and equipment.

(6) Conducts physical conditioning program.

(7) Conducts briefings, notifies Pistol Division Personnel of appointments, duties and administrative requirements.

(8) Attends to administrative requests of Pistol Division personnel.

(9) Participates in registered competition and practical training activity.

e. LINE COACH, PISTOL DIVISION:

(1) Supervises the physical and mental preparation of members of the team. This function includes assignment of target, relay and additional duties in support of team effort.

(2) Guides and influences the performance of the shooter on the firing line.

(a) Stimulates morale, attitudes, enthusiasm and confidence of team members.

(b) Develops and improves coaching technique and recommends new training ideas to Head Coach.

(c) Evaluates current performance and potential of each team member so as to be able to advise Head Coach on any necessary changes in composition of team.

(3) Responsible for posting of team scores on MTU Master Score board during practice firing and during registered matches.

(4) To be able to remain with and maintain control of his team on the firing line during matches, the coach is required to thoroughly brief the designated scorer that he furnishes to score for the adjacent team. This may well be one of his off-relay team members if there are insufficient support personnel available. The team captain is responsible for supervising the scoring of his team which is generally accomplished by a representative from an adjacent team.

(5) Assists in preparation of instructor courses and is a principal instructor in these courses.

(6) Participates in individual portion of registered competition.

(7) Is available for advice during periods between individual matches.

(8) Continuous checking of equipment and weapons for proper functioning, cleanliness, accuracy, security and safety in handling.

(9) Supervise by direct observation during individual matches, the scoring responsibility of individual shooters. Remind shooters to post scores on MTU Master Score board.

(10) Observe and intercede if an argument, protest or an infraction of the rules concerns an Army team member.

f. INSTRUCTOR SHOOTER, PISTOL DIVISION:

(1) The shooter represents the US Army in all competition.

(2) Instruct and coach shooters of less experience and ability during instructor training courses.

(3) Continuously check weapons and equipment for proper functioning, cleanliness, accuracy, security and safe handling. NCOIC must be notified of all exchanges, drawing or turn-in of any issue weapons.

(4) Responsible for scoring adjacent team member or competitor during practice and matches.

(5) Actively coach a team member on firing line in the absence of the regular coach.

g. ARMORER OR GUNSMITH, PISTOL DIVISION:

(1) Responsible for the proper mechanical functioning, operative safety devices inherent accuracy, repair and certain preventative maintenance checks of weapons.

(2) Research and development of accuracy and dependability of weapons and ammunition so as to remain abreast of improved techniques in the use of weapons.

h. RANGE SERGEANT, PISTOL DIVISION:

(1) Operates Pistol Range and controls all range firing.

(2) Issues all ammunition for training and competition.

(3) Controls issue of all supplies during trips to matches.

(4) NCOIC of transportation during trips to matches.

(5) Assist in conduct of periodic serial number checks of all issue weapons.

i. CLERK, PISTOL DIVISION:

(1) Assistant to Executive officer in administrative functions.

(2) Maintains file and does all typing.

(3) Assistant to Range Sergeant in issue of supplies and ammunition on trips.

(4) Assistant driver on trips to matches.

3. ADMINISTRATIVE REQUIREMENTS.

a. At home station.

(1) All TDY personnel must have personnel, pay and medical records.

(2) Regular pay and travel pay must be brought up-to-date.

(3) E. T. S. during training period.

(4) NRA membership check for all pistol team personnel.

(5) All personnel must have proper travel orders to cover remaining travel. (Turn-in copy of orders on arrival, to Sergeant Major.)

(6) Eye examinations.

(7) If shooter's family is present, quarters address.

(8) Check for adequate billets (Senior resident NCO is charge).

(9) Check for adequate mess facilities.

(10) Register personal weapons with Provost Marshal.

(11) Register automobile with Provost Marshal.

(12) Orient shooters on security of weapons. Use of unit storage facilities permitted.

(13) Sick call.

(14) Mailing address - locally and on trips.

(15) Mail call.

(16) Unit Bulletin Board.

(17) P.I.O. forms and pictures.

(18) Traffic Law Enforcement.

(19) Punctuality at formations.

(20) Leaves and passes.

b. Away from home station.

(1) Training Schedule - Headcoach.

(2) Strip Map - Executive Officer.

(3) Armorer will be present on range during all match firing and immediately available on firing line during team matches - NCOIC.

(4) Register Billets - NCOIC.

(5) Issue Ammunition - Range/Supply Sergeant.

(6) After Action Report Information - Executive Officer.

(7) Weekly Report to S3 - Executive Officer.

(8) Match Programs - Executive Officer.

(9) Squadding Tickets - NCOIC.

(10) Advance Travel Pay - Executive Officer.

(11) Travel Orders - Executive Officer.

(12) Sign Out and Sign In times - NCOIC.

(13) Truck Loading List - Executive Officer.

(14) Pick-up Team Awards after completion of match - Range/Supply Sergeant.

(15) Shooter's Equipment - NCOIC.

(16) Invoice to Match Officials - Executive Officer.

(17) Lunches for Enlisted Personnel - NCOIC.

(18) Proper Police of Government Billets - NCOIC.

(19) Personnel Notice of Emergencies - NCOIC.

(20) Make Team Entries - NCOIC.

(21) Have triggers weighed on all weapons, periodically and immediately prior to any match - NCOIC.

(22) Arrange for government transportation to and from range - NCOIC.

(23) Use same lot number of ammunition in match competition as used during practice firing - Range/Supply Sergeant.

(24) Competitor numbers recorded on master score board before match shooting starts - NCOIC.

(25) Each team's assigned target numbers for team matches will be posted on team score sheets located on master score board before team match starts - NCOIC.

(26) Each firing member of each team will have his relay assignment on team score sheet before team match starts - NCOIC.

(27) Shooters are personally responsible for transportation, billets and finances while on trips.

(28) Prompt departure and arrival time will conform to designated Sign Out and Sign In time.

(29) Shooter's name and all passenger's names be filled in on car card which requires description and tag number of automobile. Turn-in to NCOIC.

(30) All personnel will follow route designated on strip map unless otherwise directed.

(31) All personnel will furnish NCOIC with address of billet and telephone number if match location requires off-post billets.

(32) Personnel desiring leaves or delay en route at completion of TDY must have consent from parent unit.

(33) Reservations for family billets at any match locality including National Matches at Camp Perry is the responsibility of the individual concerned.

(34) A report of match attendance by category and total (Executive Officer.)

    (a) Army personnel.

    (b) Air Force, Navy, Marines, Coast Guard etc.

    (c) Reserve components.

    (d) National Guard.

    (e) Police.

    (f) Civilian.

(35) A report of match results of each competition participated in to include team aggregate standings.

All administrative burdens must be handled by administrative and supervisory personnel and not by the shooter if best results are to be achieved. The most routine and simple administrative concern is a measurable distration to the shooter.

ANNEX TO TEAM ORGANIZATION AND ADMINISTRATION

(EXAMPLE) Strip Map: Columbus, Georgia to San Antonio, Texas and Return

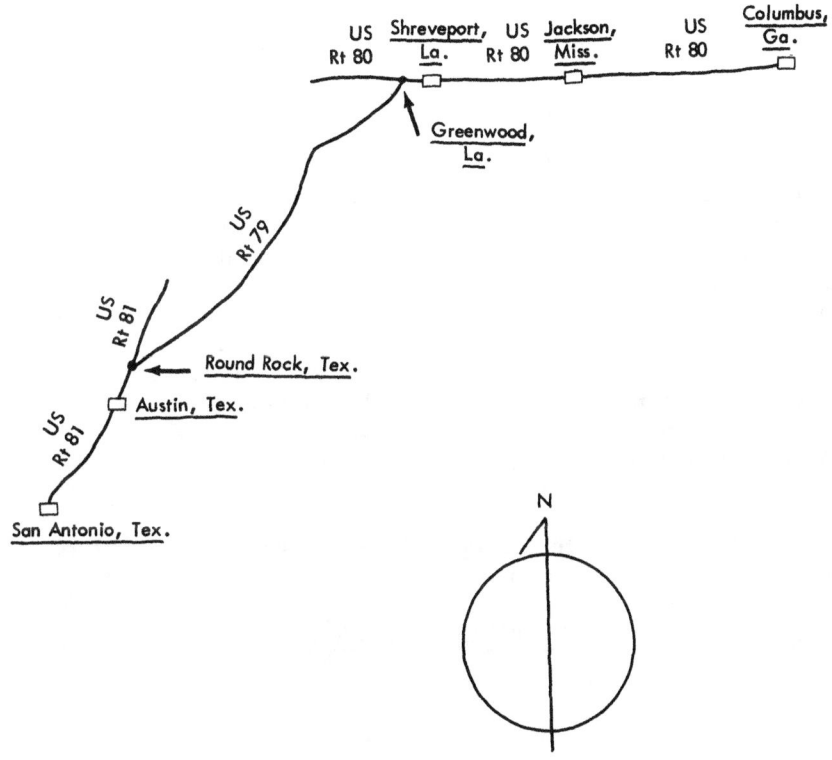

ANNEX TO TEAM ORGANIZATION AND ADMINISTRATION

SUBJECT: Clothing and equipment to be carried by the pistol shooters to matches and necessary actions to be accomplished prior to departure.

SUMMER SHOOTING UNIFORM

1. Caps, shooting (2)
2. Jacket, shooting (bush) (2)
3. Shirt, shooting (short sleeve) (3)
4. Trousers, shooting (3)
5. Boots (1 pair)
6. Belt, Web, black (1)
7. Jacket, Fatigue (1)
8. Trousers, Fatigue (1)
9. Jacket, Field (1)
10. Suit, Rain (1)

WINTER SHOOTING UNIFORM

1. Cap, shooting (1)
2. Shirts, O.G. (2)
3. Trousers, O.G. (2)
4. Jacket, field (1) with liner
5. Gloves (1 pair)
6. Belt, Web, black (1)
7. Boots (2 pair)
8. Underwear, Thermal (2 sets)
9. Jacket, Fatigue (1)
10. Trouser, Fatigue (1)
11. Jacket, shooting (bush) (2)
12. Suit, Rain (1)

CLASS "A" SUMMER

1. Insignia & Name Plate
2. Shirt, Khaki (long sleeve) (2)
3. Trouser, Khaki (2)
4. Cap, Carrison, green (1)
5. Tie, black (1)
6. Belt, Web, black (1)
7. Boots (1 pair)
8. Shoes, Low Quarter (1 pair)
9. Shirt, Poplin (1) (Officers)
10. Blouse, T.W. (1) (Officers)
11. Trousers, T.W. (1) (Officers)

CLASS "A" WINTER

1. Insignia & Name Plate
2. Uniform, Green (1)
3. Shirt, Poplin (2)
4. Cap, Garrison, green (1)
5. Shoes, Low Quarter (1 pair)
6. Boots (1 pair)
7. Belts, Web, black (1)
8. Tie, black (1)
9. Gloves
10. Overcoat

CIVILIAN CLOTHES: Appropriate and as desired. Award blazers are required attire at functions where Commanding Officer so prescribes.

ADDITIONAL ITEMS

1. Shorts
2. T-shirts
3. Socks
4. Handkerchiefs
5. Towels
6. Tags, Identification
7. Raincoat
8. Alarm clock
9. Toilet Articles
10. Shoe shining equipment
11. Brass shining equipment
12. NRA card, Identification Card, Drivers' License
13. Stationery, stamps, pen, pencil, etc.
14. Lock, keys
15. Extra coat hangers

SHOOTING EQUIPMENT

1. Gun Box, with cover
2. Guns - 22 Cal., 38 Cal., 45 Cal. W. C. & H. B.
3. Ammunition
4. Shooting Glasses
5. Scorebook & Pencil
6. Cleaning Rod and Brush
7. Cleaning Patches
8. Watch (stop)
9. Spotting Scope
10. Barrel Bushing Wrench
11. Carbide light & Carbide
12. Rosin Powder
13. Screwdriver
14. Chair or Stool
15. Extra Magazines - All Calibers
16. Binoculars (Coaches Only)
17. Sight Adjustment Card
18. Ear Plugs or Ear Protectors
19. Oil, Lubricating
20. Stapler with staples

ACCOMPLISH PRIOR TO DEPARTURE

1. Copies of Special Orders on your person and in baggage
2. Haircut
3. Safety check of automobile (lights, tires, brakes, wipers, etc.)
4. Consult strip map for route to destination.
5. SIGN OUT

REMEMBER AT ALL TIMES
CONTINUOUS MAXIMUM SECURITY OF WEAPONS

Keep your car locked and weapons in locked truck during daylight hours when weapons are not being used.

Keep weapons in quarters overnight. Do not leave weapons in unattended vehicle.

DRIVE SAFELY AND COURTEOUSLY

We can't win team matches with you in the hospital.

DON'T LET THIS HAPPEN TO YOU!

The following item from the National Safety Council Newsletter concerning what happens when an automobile collides with an immovable object, is recommended reading.

When a car hits a large tree or other immovable objects at 55 m.p.h. here is the fatal course of events that take place in seven-tenths of a second:

"First tenth of a second: The front bumper and grill work collapses, steel slivers penetrate radiator to a depth of 1 1/2 inches."

"Second tenth of a second: The hood crumples and smashes against the windshield. Fenders make contact, forcing the rear parts over the fron doors. The structural members of the car begin to act as a brake on the forward momentum, but the driver's body is still at 55 m.p.h. Legs, straight as arrows, snap at the knee joints. The noise of rending and crumpling metal is deafening."

"Third tenth of a second: Driver's body is off the seat, broken knees against dashboard. Steering wheel frame begins to bend under his grip. His head is near sun visor, chest over steering column."

"Fourth tenth of a second: First 24 inches of car is demolished. The rearend still is traveling at 35 m.p.h., the driver's body catapults along at 55 m.p.h. Motor block makes contact; rear-end of car rises from groun."

"Fifth tenth of a second: Force of forward inertia starts to impale driver on steering sheel shaft, steel punctures the lungs and arteries; blood enters lungs. The driver's body speed begins to decelerate at this point. Now only 35 m.p.h."

"Sixth tenth of a second: Driver's feet are bursting from laced shoes; brake pedal shears off at floor boards; chassis bends in middle; driver's head smashes into windshield. His chest cavity now contains the first eighteen inches of the steering column, enough to penetrate completely through."

"Seventh tenth of a second: Door latches and hinges tear loose; doors fly open. Front seat, fripped from floor, moves forward pinning impaled driver onto steering wheel column. Blood spurts from mouth; bladder bursts, shock freezes heart. The driver is dead."

In the subsequent minutes of comparative quiet, the first passersby stop and rush to the wrecked vehicle to render any aid possible and recoil in horror at the gory sight, noticing that the bloody, new statistic is sitting on his safety belt.

ANNEX TO TEAM ORGANIZATION AND ADMINISTRATION

SUBJECT: Administrative Check List

_____
NAME OF MATCH

	EXPECTED DATE	ACTUAL DATE	INITIALS
1. Program and entry cards requested on			
2. Program and entry cards received on			
3. Program and entry cards dispatched			
4. Hotel or Government Reservations			
5. Commercial Travel Reservations Confirmed			
6. Special Orders Requested			
7. Score sheets prepared on			
8. Equipment inspection prior to departure			
9. Personnel briefed on safe driving			
10. Personnel briefed on conduct and appearance			
11. Sign out time at USAMTU			
12. Sign out time at match			
13. Sign out time at match			
14. Match officials furnished sample of invoice for entry fees with instructions for preparation			
15. Arrangements made to pick up team awards			
16. Sign in time at USAMTU			
17. Preliminary Report of Match made to CO, USAMTU			
18. Equipment inspection on return			
19. Equipment lost/damaged reported to S4			
20. Attendance report of match by category			
21. After Action Report			
22. Bulletin of match results			

ANNEX TO TEAM ORGANIZATION AND ADMINISTRATION

SUBJECT: Pistol Team Vehicles, Equipment and Supplies for Pistol Match Participation.

ITEM	NR.	ITEM	NR.
Ammunition		Tacks, thumb	
Box, Gun w/spare match weapons		Tape, masking	
Binoculars, 20X		Tape, scotch	
Boxes, Tool, Pistol Armorers		Truck, Carryall	
Cans, Trash (50 gal)		Truck, 1/2-ton pickup	
Cans, Trash, Blue (5 gal)		Truck, 1 1/2-ton van	
Cans, Water, (5 gal)		Truck, 2 1/2-ton cargo	
Carbide, lb		Truck, 10-ton semi trlr van	
Cards, POL Credit		Typewriter	
Chairs, folding, aluminum		Umbrella, beach	
Cleaner, bore		Oil, lubricating	
Clipboards		Paper, bond, ream	
Clips, paper		Paper, carbon, package	
Cups, paper, hot and cold		Pads, Writing, ruler	
Dispenser, drinking, 10-gal		Pencils, colored	
Erasers, rubber		Pencils, grease	
Guns, staple w/extra staples		Pencils, lead	
Guidons with staff		Patches, cleaning	
Ice (Purchase and obtain receipt)		Pills, Vitamin w/dispenser)	
Kit, first aid		Public address set	
Hammer, claw		Stapler, machine, desk w/staples	
Mallet, wood			
Markers, magic		Rod, cleaning, pistol	
Machine, adding, H/operated		Rags, cleaning	
Scissors		Rule books, NRA Pistol	
Stakes, steel		Rope, 100 ft	
Streamers, guidon, team colors		Stencils	
Supply records, TDY personnel		Scoresheets, individual	
Table, folding, steel		Scoresheets, team	
Table, folding, wood		Tablet, Salt (w/dispenser)	
Telescope, M49		Tissue, lens, cleaning	
Target Centers, Repair 50&25 yds		Tissue, toilet	
Manuals, Pistol		Scoreboard w/easel	
Awning, Truck			

ANNEX TO TEAM ORGANIZATION AND ADMINISTRATION

SUBJECT: Information Questionnaire for Team Members.

NAME_____ RANK_____ SN_____

Parent Organization_____

New Man (Yes) (No) National Trophy Team Match_____

Medal Winner (Yes) (No)_____

Last year you fired in National Trophy Team Match_____

Distinguished (Yes) (No)_____

Number of points toward distinguished award_____

How long since last overseas assignment?_____

Prefer (Long-Short) Trigger_____

Hat Size_____ Jacket Size_____

Trousers Waist_____ Length_____

Shirt Neck_____ Sleeve_____

ETS Date_____

Are you on levy for overseas assignment?_____

Do you have any personal problems that would preclude your staying through the complete training period?_____

Are you a member of the National Rifle Association?_____

What is your NRA Pistol Classification?_____

How many years of experience in competitive pistol shooting?_____

What is your present connection with marksmanship in your parent unit and/or home station?
_____

Do you presently hold or have previously held any NRA National Pistol records? Individual___
_____

Team_____

What is the highest score you have fired in registered NRA Match Competition:

.22 Cal Aggregate_____ Center Fire Aggregate_____

.45 Cal W.C. Aggregate_____ 45 Cal Service Ammunition Aggregate_____

Three (3) Gun 2700 Aggregate_____

ANNEX TO TEAM ORGANIZATION AND TEAM ADMINISTRATION

SUBJECT: Example of a Weekly Training Schedule

HEADQUARTERS
UNITED STATES ARMY MARKSMANSHIP TRAINING UNIT
Fort Benning, Georgia

TRAINING SCHEDULE 10 February 1965
US ARMY PISTOL DIVISION
From 13 Feb through 17 Feb 65

DATE & TIME		AREA	INSTRUCTOR	UNIFORM
Mon 13 Feb				
0800-1200	900 Aggregate, Cal. 22	Pistol Range	Head Coach	Fatigues
1300-1430	NMC Cal .22 Team Match	Pistol Range	Head Coach	Fatigues
1430-1600	NMC Cal .45 H.B.	Pistol Range	Head Coach	Fatigues
1600-1700	Physical Training	Pistol Range	Line Coach	Fatigues
Tue 14 Feb				
0800-1200	900 Aggregate, Cal .38	Pistol Range	Head Coach	Fatigues
1300-1430	NMC, Cal .38 Team Match	Pistol Range	Head Coach	Fatigues
1430-1600	NMC, Cal .45 H.B.	Pistol Range	Head Coach	Fatigues
1600-1700	Physical Training	Pistol Range	Line Coach	Fatigues
Wed 15 Feb				
0800-1200	Mandatory Training, (Map Reading)	Unit Classroom	Training Officer	Fatigues
1300-1700	Organized Athletics	Unit Area	NCO In Charge	Fatigues
Thu 16 Feb				
0800-1200	900 Aggregate, Cal. 45 W.C.	Pistol Range	Head Coach	Fatigues
1300-1430	NMC, Cal .45 W.C. Team Match	Pistol Range	Head Coach	Fatigues
1430-1600	NMC, Cal .45 H.B.	Pistol Range	Head Coach	Fatigues
1600-1700	Physical Training	Pistol Range	Line Coach	Fatigues
Fri 17 Feb				
0800-0930	NMC, .22 Cal Team Match	Pistol Range	Line Coaches	Fatigues
0930-1100	NMC, Cal .38 Team Match	Pistol Range	Line Coaches	Fatigues
1100-1200	NMC, .45 Cal W.C. Team Match	Pistol Range	Line Coaches	Fatigues
1300-1430	NMC, .45 Cal HB Team Match	Pistol Range	Line Coaches	Fatigues
1430-1530	Maintenance of Weapons	Pistol Range	Armorer	Fatigues
1530-1600	Inspection of Weapons and Equipment	Pistol Range	Shop Officer	Fatigues
1600-1700	Physical Training	Pistol Range	Line Coach	Fatigues

DISTRIBUTION "D" PEOT
  Colonel
OFFICIAL:
Wentworth
S3

ANNEX TO TEAM ORGANIZATION AND ADMINISTRATION

SUBJECT: Example of Practice scorecard (Individual)

NAME_____RANK_____DATE_____

CRS OF FIRE	SCORE		TOTALS	SCORE		TOTALS	SCORE		TOTALS	SCORE		TOTALS
SF												
TF												
RF												
NMC — S / T / R												
900 AGGREGATE												
NMC TEAM — S / T / R												
TEAM X 3 AGGREGATE												
.22 CAL AGGREGATE												
PREVIOUS AGGREGATE												
GRAND .22 CAL AGGREGATE / GRAND .38 CAL CF AGGREGATE / GRAND .45 CAL WC AGGREGATE / GRAND .45 CAL HB AGGREGATE												
STANDING												

_____   _____
SCORERS SIGNATURE   VERIFIED (FIRER) SIGNATURE

FB USAMTU FORM 8 Modified
2 FEB 1964

CHAPTER XIII

PROCUREMENT, MAINTENANCE AND SECURITY OF MATCH WEAPONS, EQUIPMENT AND AMMUNITION

The Department of the Army has made provision for organic support of a comprehensive competitive marksmanship program for all types of small arms used in the combat branches.

A. <u>SECURITY</u>. Organizational responsibility for security of weapons, equipment and ammunition is guided by directives tailored to the security requirements of local posts, camps and stations. These directives cover the responsibility of the individual to safeguard US Government property and registration of personal weapons.

B. <u>PROCUREMENT</u>. TA 60-18, November 1964 is the source of weapons and equipment. This TA lists the types and number of weapons authorized for various level teams. Items of equipment other than weapons are also listed.

    1. <u>Travel, per diem and entry fees</u> are authorized out of appropriated funds while the awards and trophies are purchased with nonappropriated funds.

    2. <u>Expendables items</u> are purchased at the self-service store. Each unit is authorized a certain amount of money for this purpose.

    3. <u>Ear plugs and prescription glasses</u> are obtained through medical channels. Medical authorities encourage wearing of ear plugs or other protective devices for the conservation of hearing.

    4. <u>Ammunition</u> allowances for various level teams are listed in TA 23-100. In this TA you will find that you can usually draw all the .22 caliber and hardball .45 that your team can shoot. Commercial .38 and .45 caliber ammunition is also authorized in limited quantities depending upon the level team concerned. Pay particular attention to the authorization for off-duty marksmanship practice for .22 caliber and .45 caliber service ammunition.

    5. <u>Match grade weapons</u> should be procured in all three calibers.

        a. Any match grade .22 caliber automatic target pistol is good. There are several good models and it is up to the individual to choose which one he prefers. A unit purchasing these weapons will usually get more than one type so as to give the shooter a choice.

        b. Center fire weapons may be any center fire pistol or revolver .32 caliber or larger. A revolver is harder to master than either a .45 caliber automatic or a .38 super automatic. The majority of shooters on the firing lines today shoot a .38 caliber super automatic. A .38 super automatic may be accurized by the unit gunsmith or they may be purchased commercially from any number of competent gunsmiths.

        c. Government issue match grade .45 caliber automatics are available in marksmanship units and most are very accurate.

        d. Commercial .45 caliber automatics may be purchased and accurized by the various gunsmiths.

    6. <u>Miscellaneous Equipment.</u> Any number of gun boxes are available and one of good solid construction should be chosen. Spotting scopes M49 are available for issue and are of excellent quality. Stop watches are available for issue and other miscellaneous items such as scoring plugs, carbide lamps, etc., are available either from TA 60-18 or local purchase.

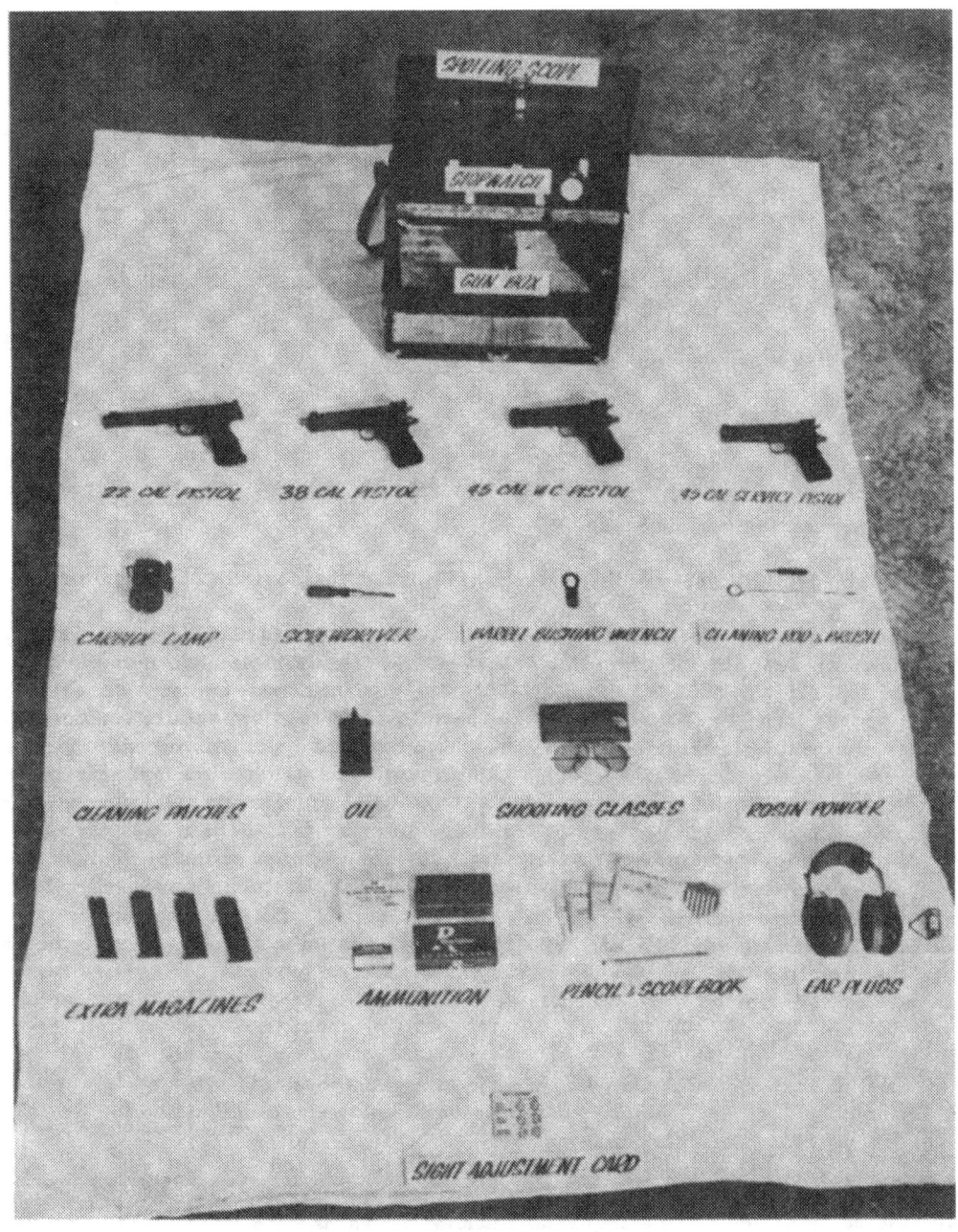

Figure 1

C. CAUSES OF WEAPONS MALFUNCTIONS AND CARE AND MAINTENANCE. Weapons used in competition are of necessity closely fitted. Occasionally a weapon will malfunction until it has been fired a few hundred rounds. Some magazines may not work in a weapon due to the lips being either too wide or too close together. The followers may be at an improper angle, causing the rounds to either ride too high and hit the upper edge of the chamber or not high enough in which case it may not properly slide up the feed ramp and into the chamber. Magazines should be handled carefully so that they do not become bent and they should be taken apart and cleaned occasionally. All wearing surfaces of the weapon and the magazines should be well oiled at all times with oil preservative light, as issued. Lack of lubrication cause excess wear, malfunctions and materially reduces the accuracy life, making it necessary to reaccurize the weapon. Cleaning materials as issued are adequate and should be used daily after firing. A dirty weapon also causes stoppages and excessive wear. When transporting weapons they should be fastened securely within the box so as not to damage the sights. It is mandatory that weapons be locked up at all times when not in use and should never be left unattended.

D. <u>THE MARKSMANSHIP UNIT SHOP IS PART OF THE TEAM EFFORT</u>. Each Marksmanship Unit should have a shop where the unit gunsmith has facilities to:

1. Accurize weapons.

2. Test weapons for inherent accuracy.

3. Make major repairs.

4. Do preventive maintenance.

5. Make special adjustments on weapons to meet the desires of individual shooters.

A machine rest or static testing device is essential for testing of newly accurized weapons and for checking weapons to determine if they need reaccurization after extensive use. The shop is a very important part of the team effort. The quality of work turned out is directly reflected in the scores fired by the team. Target pistols accurized especially for competitive shooting undergo careful checking and adjustment of parts and mechanisms under precise conditions. However, they may require special adjustment, to a greater or lesser extent. The tooling to fit and the adjustment of the pistol, which is intended to improve the interaction of the parts and mechanisms, as carried out by armorers, fails in many instances to completely meet the individual shooter's requirements. For example in the adjustment of the trigger, the altering of the sighting devices to suit the shooter's sharpness of vision, etc. It is necessary for the shooter to inform the armorer just how he wants the weapon adjusted. This responsibility is the shooter's and cannot be delegated to anyone else. For this reason, when a pistol is sent to the shop for reaccurizing, the pistol should be reissued to the same shooter because he is more than likely delicately attuned and coordinated to the trigger action and the operating rate of that particular weapon.

6. All match grade ammunition used in competition must be tested for uniform accuracy by lot number. Match weapons should be fired using the ammunition lot that gives the greatest accuracy to the particular weapon.

7. Shot group areas of dispersion.

If it were possible to fire a series of rounds under identical conditions, the bullets would describe the same trajectory in the air and strike the same point.

However, in practice it is almost impossible to have absolute uniformity of all firing conditions, since there are always small, practically indiscernible variations in size of powder grains, weight of charge and bullet, shape of bullet; different igniting ability of primer; various conditions of bullet motion in the barrel and outside it, such as constant fouling and heating of the barrel, wind gust and changing air temperature; errors committed by the shooter in aiming, assuming the firing position, etc. As a result, even under the most favorable firing conditions, each bullet describes a trajectory which differs from the trajectories of other bullets. This phenomenon is called <u>dispersion</u>.

If a sufficiently large number of rounds are fired, the trajectories form a sheaf of trajectories, which produces a number of shot holes separated by various distances in the target. The area the holes occupy is called the area of dispersion.

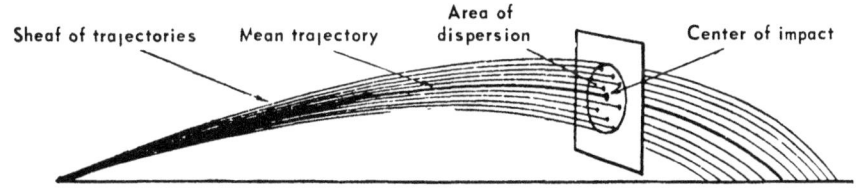

Figure 2. Sheaf of Trajectories. Mean Trajectory. Area of Dispersion.

Without going into detail to explain the laws of dispersion and the concepts of the probability of hitting, let us say that in shooting any type of firearm, the character of the shot pattern in the area of dispersion always remains the same, even though the size of the areas of dispersion can differ to a great extent, depending upon the model of weapon and ammunition used.

It follows from the above that dispersion is an objective process occurring independently of the will and desires of the shooter. It is senseless to require all bullets to strike the same point.

However, dispersion is not an unavoidable and invariable amount which is definitely fixed for a certain weapon under a definite firing condition. The art of accurate shooting and gun smithing success lies in recognizing the causes of dispersion and diminishing their effect.

One of the actions that can be taken by the individual shooter to reduce dispersion, in addition to the normal mechanical fitting of the pistol components by the gunsmith, is the maintenance of a clean, unleaded pistol bore. Lead begins to build up in .38 Cal., and .45 Cal. Pistol barrels after twenty (20) to thirty (30) rounds fired. It is recommended that a brass wire brush be swabbed through the bore after each ten (10) rounds fired so as to remove the small initial lead accumulations before a larger build-up effects the accuracy of the pistol. A suitable solvent should be used to thoroughly clean and remove all lead fouling after the completion of the days firing.

Marksmanship makes very high demands upon a gun's accuracy of fire and upon the ammunition. No matter how well trained and capable the shooter may be technically, his competitive results will depend to a large extent upon the quality of the gun and the ammunition.

SECTION FOUR
COMPETITIVE REGULATIONS

CHAPTER XIV

NRA PISTOL MATCH RULES

NRA Pistol Match Rules cover completely the conduct of all registered NRA Pistol Matches Current NRA pistol match rule books are available from NRA Headquarters, 1600 Rhode I. St. NE, Washington 6, D.C., or any NRA official referee.

In order to play the game properly, all the players should know the rules. Ignorance of the rules of pistol competition may jeopardize the outcome of any match. Knowledge of the rules avoids the pitfalls that may result in disqualification of a team score.

Only those rules and regulations most frequently used will be covered briefly in this discussion. For specific cases, references should be made to a current issue of the NRA pistol rule book.

A. <u>TYPES OF COMPETITION</u> - (RULE 1)

There are approximately eleven popularly recognized types of competition.

    1. (1.2) Open Match - A match open to anyone. An open match may be limited to citizens of the United States or to members of the National Rifle Association of America. Such limitation must be stated in the program.

    2. (1.2.1) National Trophy Matches - The National Trophy Matches are organized and conducted under the direction of the National Board for the Promotion of Rifle Practice (NBPRP). These matches are conducted in accordance with rules and regulations contained in the Army Regulations 920-30.

    3. (1.2.2) National Matches - The National Matches are the combined NRA National Championships and the National Trophy Matches.

    4. (1.3) Restricted Match - A match in which competition is limited to specified groups, i.e., juniors, women, police, civilians, veterans, etc.; or to specified class, i.e., Master, Expert, Sharpshooters, Marksmen, etc.

    5. (1.4) Classified Match - A match in which prizes are awarded to the winners and to the highest competitors in several specified classes, such as Masters, Experts, Sharpshooters, Marksmen. The classification of competitors may be accomplished by the National Classification System (Sec. 19) or by other means. The program for classified matches must specify the groups or classes in which awards will be made.

    6. (1.5) Invitational Match - A match in which participation is limited to those who have been invited to compete.

    7. (1.6) NRA Competition - Competition sanctioned in advance of firing by the National Rifle Association. The program, range facilities and officials must comply with standards established by the NRA. (See Section 21.)

    8. (1.7) League Competition - A form of competition in which teams compete one against another under a pre-arranged schedule in a series of matches. Leagues usually provide for each team to fire against each other team at least once during the league season. Final standings are usually determined by the percentage of matches won by those who fired the required number of matches. Special prizes may be awarded for high individual or team scores.

9. (1.8) Squadded Individual Match - A match in which each competitor is assigned a definite relay and target by the statistical office. Failure to report on the proper relay or firing point forfeits the right to fire. All entries must be made before firing commences in that match, except when otherwise stated in the tournament program. (Rule 9.20)

10. (1.11) Squadded Team Match - A match in which the teams are assigned a definite time to fire. Teams may be assigned one or more adjacent targets. All entries must be made before firing commences in that match. The entire team must report and fire as a unit. (See Rule 9.20)

11. (1.13) Aggregate Matches - An aggregate of the scores from two or more matches This may be an aggregate of match stages, individual matches, team matches, or both, provided the tournament program clearly states the matches which will comprise the aggregate. Entries in aggregate matches must be made before the competitor commences firing in any of the matches making up the aggregate match.

12. (1.13.1) Tournament - A tournament is a series of matches covered by an official program. Such matches may be all individual matches, all team matches or a combination of both; they may be all fired matches or a combination of fired and aggregate matches. A tournament may be conducted on one day, on successive days or may provide for intervening days between portions of the tournaments programmed for conduct over more than one weekend.

B. <u>COMPETITIVE REGULATIONS</u> - (RULE 9)

The numerous competitive regulations affecting the conduct of a pistol match are designed to promote safe, efficient operation of the range and thereby preclude interruptions, delays and confusion in the progress of a pistol match.

1. (9.1) Pistols Unloaded - Pistols will not be loaded until the competitor has taken position at his firing point and the command "Load" has been given.

2. (9.2) Actions Open - Unless Pistols are holstered or cased, cylinders must be open or slides back at all times until the competitor is in position at his firing point and the command "Load" has been given.

3. (9.3) Pistols Loaded - No pistol will be loaded until competitor has taken his place at the firing point and command "Load" has been given by the range officer. Loaded pistols shall be pointed in the direction of the targets at all times.

4. (9.4) Cease Firing - All pistols will be unloaded immediately upon the command "Cease Firing." Actions will remain open.

5. (9.5) Not Ready - It is the duty of competitors to properly notify the range officer if not ready to fire at the time the command is given, "Is the line ready?" Should the Chief Range Officer cause firing to proceed, the competitor concerned will be given an opportunity to fire his score in the earliest possible relay. Failure of competitor to notify the range officer that he is not ready forfeits his right to fire.

6. (9.6) Changing Pistols - No competitor will change his pistol during the firing of any match (except aggregate matches), unless it has become disabled and has been so designated by the Chief Range Officer. Claim that a pistol is disabled must be made immediately. All shots fired up to the time the claim is made will stand as part of the official score. The exchange of barrels, portable weights, etc., shall not be restricted. (For timed and rapid fire see 9.10, 9.24 and 9.27)

7. (9.9) Defective Cartridge - Is one (a) that is unsafe to fire by reason of improper loading or structural deficiencies, (b) that fails to fire when the primer is indented by the firing pin, (c) from which the bullet has not left the barrel. No claim for defective cartridge will be allowed if bullet has left the barrel. (For procedure in case of a defective cartridge, see Rule 9.23 and 9.24. For refiring privileges see Rule 9.27)

8. (9.10) Disabled Pistol - Any pistol which cannot be safely aimed or fired, or has suffered the loss of a sight or damage to the sights rendering it impossible to properly aim at target or cannot be fired because of mechanical failure. There must be evidence of physical damage to sights, the fact that sights are improperly adjusted does not constitute disablement. A pistol once declared disabled by the range officer shall not again be used for competitive firing until the defect has been corrected and until the pistol has been ruled as safe by the Chief Range Officer. No competitor will be allowed more than one refire per match because of a disabled pistol, defective cartridge or malfunction. (For procedure in case of a disabled pistol see Rules 9.23 and 9.24. For refiring privileges see Rule 9.27)

9. (9.11) Malfunction - Failure of the pistol to function properly due to mechanical defects or to defective ammunition. Functional failures due to improper manual operation are not to be considered as malfunctions. (For procedure in case of a malfunction see Rules 9.23 and 9.24. For refiring privileges see Rule 9.27)

10. (9.14) Weighing Triggers - Triggers may be weighed with official NRA trigger test weights, at the discretion of the Executive Officer, Official Referee or Supervisor. Failure of the trigger to meet the trigger pull requirements shall disqualify the competitor in matches previously fired. While trigger pull is being weighed, the pistol shall be held with the barrel perpendicular to the horizontal surface on which test weight is supported. The rod or hook of test weight shall rest on lowest point of the curve in curved triggers, or on a point approximately one quarter of an inch from lower end of straight triggers. To pass the weight test, a weight of the correct number of pounds shall be lifted by the pistol trigger while in the cocked position and while all safety devices are in firing position from the horizontal surface on which it is resting, until the weight hangs free and without releasing the trigger. Magazine must be removed and pistols unloaded while trigger is being weighed. Pistols equipped with a device to prevent firing while magazine is out must be closely inspected to see that no cartridges are in magazine or chamber. Magazine will then be inserted and trigger pull weighed. Competitors will be permitted to adjust triggers which have failed to pass the weight test provided they do not occasion any delay. Failure of trigger to pass the weight test is the competitor's responsibility.

11. (9.16) Competitors Position - Competitors will take their position at their numbered firing point in such manner so as to not interfere with competitors on either side. No portion of the shooter's body may rest or touch the ground in advance of the firing line.

12. (9.17) Individual Coaching - Coaching is prohibited in all individual matches unless otherwise provided in the conditions of the match.

13. (9.22) Interruption of Fire -

    a. In timed or rapid fire when the firing of a string is interrupted by some occurrence which renders it impossible for one or more competitors to complete the string under the conditions of the match, the Chief Range Officer will proceed as follows. Without being permitted to examine their targets, competitors in the relay who have been so prevented from completing their strings will be asked if they wish to refire or to accept their score as fired. Targets will then be scored in the usual manner for all competitors except those who have elected to refire. Without being scored the targets of such competitors who have elected to refire will be pasted or new targets substituted and a complete string will be fired and scored. Reasons authorizing this procedure are (1) failure to allow full time, (2) failure of targets to operate properly or uniformly for the entire string, (3) failure of paper target to remain in position on frame or carrier, (4) damage to target rendering impossible proper aiming or scoring, (5) the appearance of some object in line of fire constituting a hazard, (6) some accident involving a Range Officer or competitor on the firing line, (7) unintended moving of the target during firing.

      b. In timed or rapid fire when, due to faulty target operation or error in timing, one or more competitors are allowed more time to complete the string than provided by conditions of the match. The Chief Range Officer will immediately order all such targets pasted or new targets installed. The fired targets will not be scored. A complete new string will then be fired by the competitors who were allowed excess time. If in the same relay some targets operate properly in accordance with the legal time limit, such targets will be scored in the usual manner and competitors firing on those targets will not be required or permitted to refire.

      c. In slow fire in case target is unintentionally moved out of firing position just as a shot is fired the shot will be ringed by the Pit or Range Officer if it can be identified, and disregarded when target is scored. If the shot cannot be identified, the competitor will be permitted to accept the score as fired or to fire a completely new string.

      d. In case of excessive hits see Rule 14.10.

      e. If a shot hits the target frame or the target carrier causing the target to fall, the range officer will be notified. The target will be rehung and if the shot causing the target to fall strikes outside the scoring rings of the target it will be scored as a miss. All other shots on the target will count as record shots. For slow fire the competitor will be permitted to continue to fire any unfired shots. For timed and rapid fire the competitor will refire the string on the same target as provided in Rule 9.24.

   14. (9.23) Failure to Function in Slow Fire - If a cartridge fails to fire or a pistol fails to function in slow fire the competitor will call the range officer. The range officer, when satisfied that there is a defective cartridge (Rule 9.9), disabled pistol (Rule 9.10) or malfunction (Rule 9.11) will permit the competitor to replace the unfired cartridge or clear the jam and continue firing. Additional time may be allowed each competitor, equal to the time lost because of the defective cartridge, disabled pistol or malfunction. (For refiring privileges see Rule 9.27 b)

   15. (9.24) Failure to Function in Timed or Rapid Fire -

      a. In the event of a defective cartridge (Rule 9.9), disabled pistol (Rule 9.10) or malfunction (Rule 9.11), before a string is completed in timed or rapid fire the competitor shall be privileged to fire another five shot string, provided he assumes the ready position and calls the Range Officer by holding up his hand at the end of the time period. The Range Officer will inspect the pistol, and if satisfied that there is a defective pistol, defective ammunition, or malfunction, will determine the number of unfired cartridges remaining in the pistol, or bullets that have failed to leave the barrel. The competitor will then fire another complete five shot string on the same target.

      When scoring is after each five shot string - the competitor will be charged with firing ten shots less whatever number of cartridges were found unfired or bullets have failed to leave the barrel in the original string and will be scored the five shots of lowest value.

When scoring after ten shots - the competitor will be charged with firing fifteen shots less whatever number of cartridges were found unfired or bullets that failed to leave the barrel in the original string and will be scored the ten shots of lowest value.

Unfired shots in the refire string will be scored as misses. Failure of the competitor to notify the Range Officer of the malfunction, or the opening or clearing or attempting to clear the pistol by the competitor before the Range Officer has inspected the pistol forfeits the right of the competitor to refire. (For refiring privileges see Rule 9.27 c)

16. (9.25) Interference with Target - Competitors will not be permitted to interfere with the handling of targets by range personnel and competitors will not be permitted in the pits except when assigned there as pit detail. No competitor shall touch his own target after it has been fired at until final score determination on the target has been made. Final score determination is not reached until all challenges have been settled.

17. (9.29) Score and Classification Falsification - No competitor will falsify his score or classification, nor that of any other competitor, nor be an accessory thereto.

18. (9.32) Disorderly Conduct - Disorderly conduct or intoxication is not permitted on the range and anyone guilty of same will be expelled from the range.

19. (9.33) Refusal to Obey - No person will refuse to obey instructions of the Official Referee, Supervisor, Executive Officer, Range Officer or any other officer of the tournament in any instructions given in the proper conduct of his office.

20. (9.34) Evasion of Rules - No competitor will evade nor attempt to evade nor be an accessory to the evasion of any of the conditions of a match as prescribed in the program or in these rules. Refusal of a competitor or tournament official to give testimony regarding facts known to him concerning violations or attempted violations of these rules will constitute being an accessory to the violation or attempted violation.

21. (9.35) Disqualification - The Official Referee, Supervisor or Executive Officer upon proper presentation of evidence will disqualify any competitor for violation of any rule or parts of Rules 9.25 to 9.34 inclusive. They may disqualify any competitor or order his expulsion from the range for violation of any other of these rules or for any conduct they consider discreditable.

22. (9.36) Suspension - The Executive Committee or Protest Committee of the National Rifle Association upon presentation of evidence and hearings as provided in the Association's bylaws may suspend or expel any member for violation of these rules.

C. <u>CHALLENGES AND PROTESTS</u> - (RULE 16)

Challenge and protest procedures respectively apply to competition wherein a competitor feels that a shot has been improperly evaluated or feels that an injustice has been done to him other than in evaluation of the target.

1. (16.1) Challenges - When a competitor feels that a shot fired by himself or by another competitor has been improperly evaluated or scored, he may challenge the scoring. Such challenge must be made immediately upon announcement of the score. No challenge will be accepted after target has been pasted.

    a. A challenge fee of not more than $1.00 may be charged to all competitors making challenges. The challenge fee will be collected before making the first re-check of the challenged score. If the competitor's challenge is sustained at any point along the line of rechecks, the challenge fee will be returned to him. If the challenge is lost, the challenge fee will be included in the general revenue of the tournament. In NRA Competition to which the NRA assigns a Referee or Supervisor the decision of the Official Referee or Supervisor will be final except in the National Championships.

b. When targets are scored on frame and scoring of a shot is challenged the Range Officer will immediately call the Official Referee or Supervisor who will score the target. If necessary to avoid delaying match the challenged target will be replaced with a clean target and the match will proceed. Official Referee or Supervisor will score the target as soon as possible and notify the competitor.

2. (16.2) Protests - A competitor may formally protest:

a. Any injustice which he feels has been done him except the evaluation of a target, which he may challenge as outlined in Rule 16.1.

b. The conditions under which another competitor has been permitted to fire.

c. The equipment which another competitor has been permitted to use.

3. (16.3) How to Protest - A protest must be initiated immediately upon the occurrence of protested incident. Failure to comply with the following procedure will automatically void the protest:

a. State the complaint verbally to Chief Range (Chief Statistical) Officer. If not satisfied with his decision then,

b. File with the Official Referee or Supervisor a formal protest in writing stating all the facts in the case. Such protest must be filed within twelve hours of the occurrence of protested incident. If not satisfied with decision of Official Referee and Supervisor then,

c. File with the Executive Committee of the National Rifle Association a written appeal stating all facts. Such appeal must be filed with the NRA Official Referee or Supervisor within 12 hours after his decision has been made known to the competitor. This protest will then be forwarded to the NRA office by the Official Referee or Supervisor with a complete statement of facts, within 12 hours of the time it is filed with him.

4. (16.4) Team Matches - Must be made by the Team Captain. Team members who believe they have reason to challenge or protest will state the facts to their Team Captain who will make the official challenge or protest if he feels such action is justified.

D. EQUIPMENT AND AMMUNITION - (RULE 3)

It is not the intention of NRA rules to restrict the legal use of weapons, equipment or ammunition. Any restrictions or conditions desired must be stated by the match sponsor in the match program.

1. (3.1) Service Pistol Cal .45 - U.S. Pistol, caliber .45 M1911 or M1911A1 or the same type and caliber of commercially manufactured pistol. The pistol must be equipped with issue or similar factory standard stock (i.e., without thumb rests). Trigger pull must be not less than 4 pounds. The pistol must be equipped with open sights. The front sight must be non-adjustable. The pistol may be equipped with an adjustable rear sight with open U or rectangular notch, the distance between sights measuring not more than 7 inches from the apex of the front sight to the rear face of the rear sight. The forestrap of the grip may be checkered. The mainspring housing may be either the flat or arched type. Trigger shoes

may be used. Trigger stops, internal or external, are acceptable. Otherwise, external alterations or additions to the arm will not be allowed. The internal parts of the pistol may be specially fitted and include alterations which will improve the functioning and accuracy of the arm, provided such alterations in no way interfere with the proper functioning of the safety devices as manufactured. All standard safety features of the weapon must operate properly. It is the competitor's responsibility to have his weapon checked prior to the firing of the match.

2. (3.2) Pistol Cal .45 (Wad Cutter) - Any .45 caliber semi-automatic pistol, trigger pull not less than 3 1/2 pounds, sights may be adjustable but not more than ten inches apart. All standard safety features of weapon must operate properly. (EXPLANATORY Note: Any sights including telescopic are permitted.)

3. (3.3) Service Revolver .45 - The revolver, caliber .45, model 1917, with stocks and sights as issued, or of identical dimensions and designs, except that stocks and backstrap may be checkered. Trigger pull not less than 2 1/2 pounds.

4. (3.4) Any Center Fire Pistol or Revolver - Center-fire pistols (single shot or semi-automatic) or revolver of .32 caliber or larger (including 7.65mm and .45 caliber pistols and revolvers); barrel length, including cylinder, not more than ten inches; trigger pull not less than 2 1/2 pounds, except .45 caliber semi-automatic pistols not less than 3 1/2 pounds. Sights may be adjustable but not over ten inches apart. All standard safety features of weapon must operate properly. Programs may specify particular calibers or types of center-fire weapons which will be permitted or not permitted in stated event. (EXPLANATORY NOTE: Any sights including telescopic are permitted.)

5. (3.5) .22 Caliber Pistol or Revolver - Any pistol (single shot or semi-automatic) or revolver using a .22 caliber rim-fire cartridge having an over-all length of not more than 1.1 inches and with lead or alloy bullet not greater than .23" in diameter and weighing not more than 40 grains; barrel length, including cyclinder, not more than ten inches; sights may be adjustable but not over ten inches apart trigger pull not less than 2 pounds. (EXPLANATORY NOTE: Any sights including telescopic are permitted.)

6. (3.8) Spotting Scopes - The use of the telescope to spot shots is permitted.

7. (3.9) Shooting Kits - The shooting kit may be taken to the firing point when it is of such size and construction as to not interfere with shooters on adjacent firing points.

8. (3.9.1) Deflecting Screens - Shooters may use a screen fastened to their pistol kit to deflect empty cases; made of four mesh, 20 gauge, not to exceed 18 gauge. Solid materials may not be used.

9. (3.17) Ammunition -

a. Service-Full charge ball cartridge ammunition manufactured for or by the Government and issued by the United States Army for use in service arms.

b. Any-Ammunition of any description that may be fired without danger to competitors, range personnel or equipment. Tracer, incendiary, armor piercing and similar ammunition is prohibited.

E. <u>SCORING</u> - (RULE 14)

Targets are scored by an adjacent competitor or a person designated by the range officer. All participants in pistol competition will be called upon to act as scorers.

1. (14.1) When to Score - Targets are scored after each ten(10) shot string or each five (5) shot string.

2. (14.2) Where to Score - Targets may be scored on the target frames, at the firing line, in the Statistical Office or in the target pit. The scorer must be at the target when scoring.

3. (14.3) How to Score - A shot hole, the leaded edge of which comes in contact with the outside of the bull's-eye or scoring rings of a target, is given the higher value. A scoring gauge approved by the NRA will be used to determine the value of close shots. The higher value will be allowed in those cases where the flange on the approved gauge touches the scoring ring. No scoring gauge will be used unless the diameter of the scoring flange is within these limits:

.22 caliber, .2225"-.224"
.32 caliber, .310"-.314"
.38 caliber, .355"-.359"
.44 caliber, .426"-.430"
.45 caliber, .450"-.454"

SCORING (RULE 14)

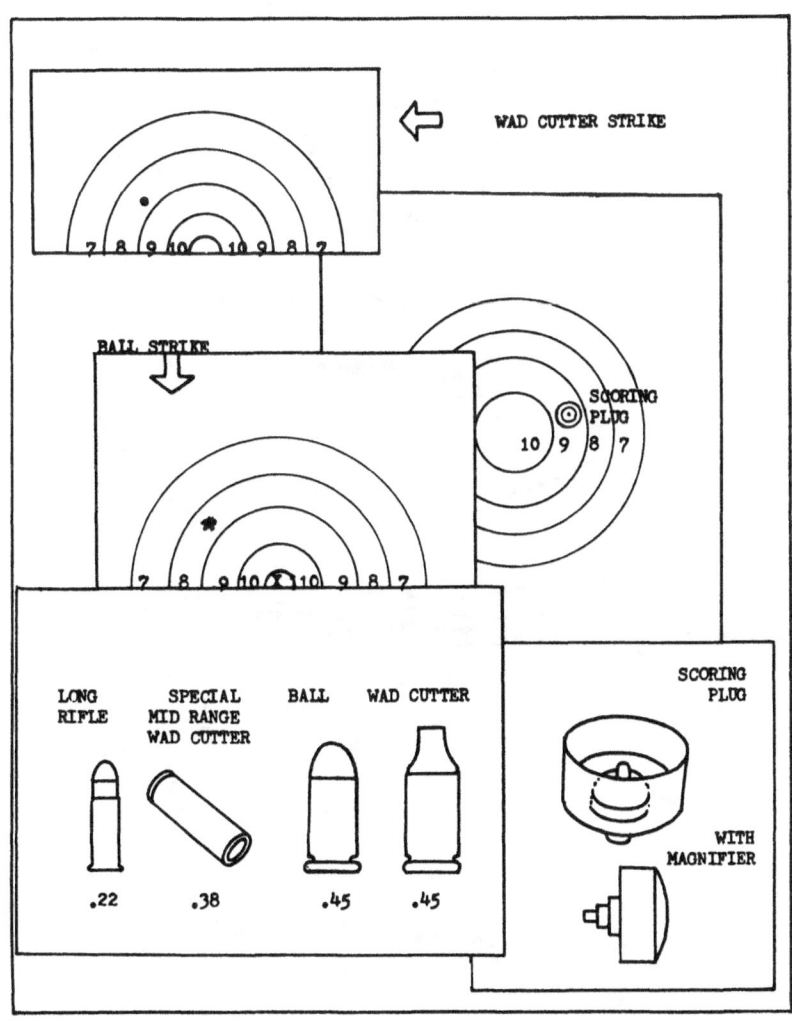

Figure 1. Bullet Strikes, Types of Ammunition and Scoring Plugs.

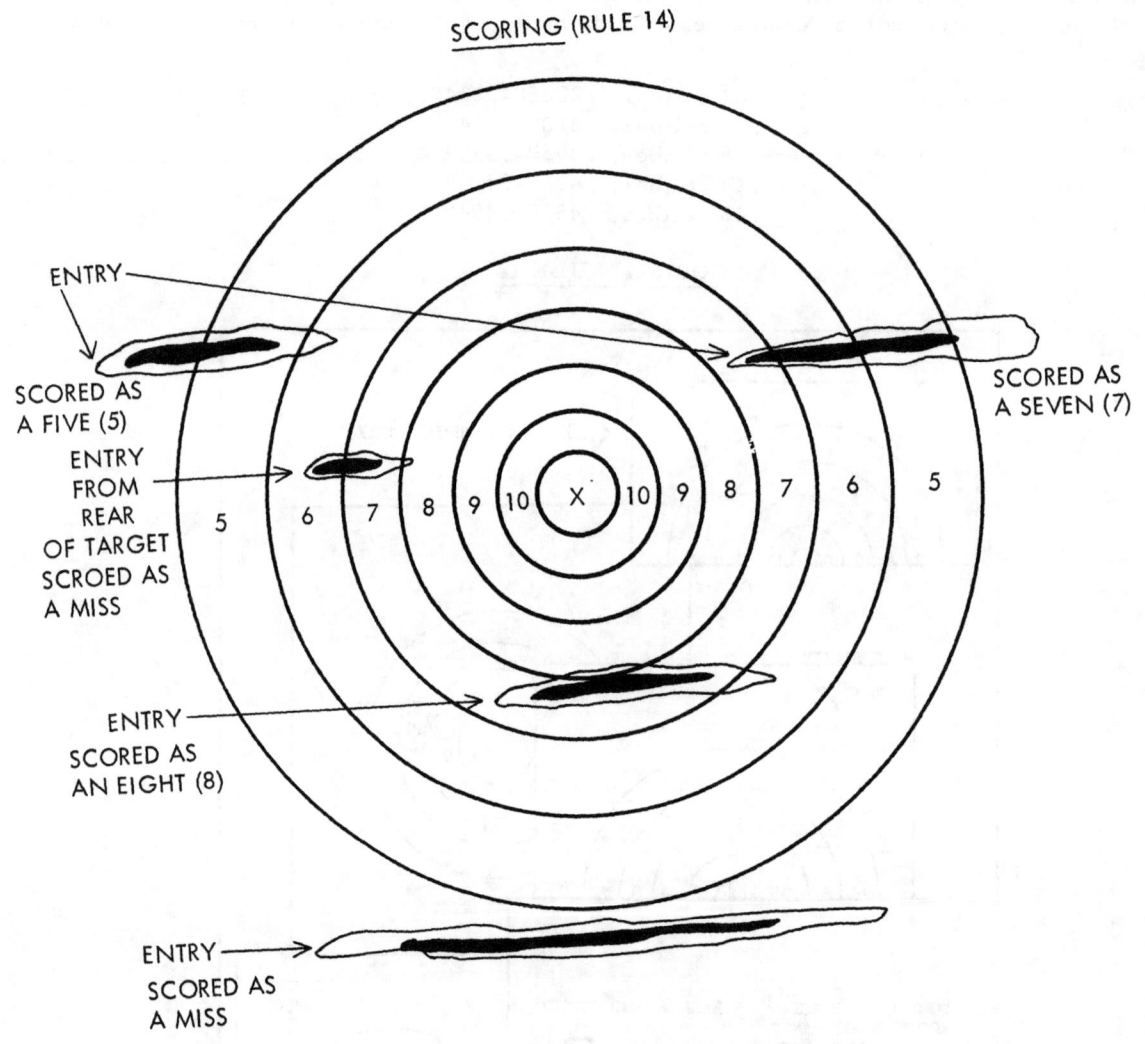

Figure 2. Timed and Rapid Fire Skidders.

SCORING (RULE 14)

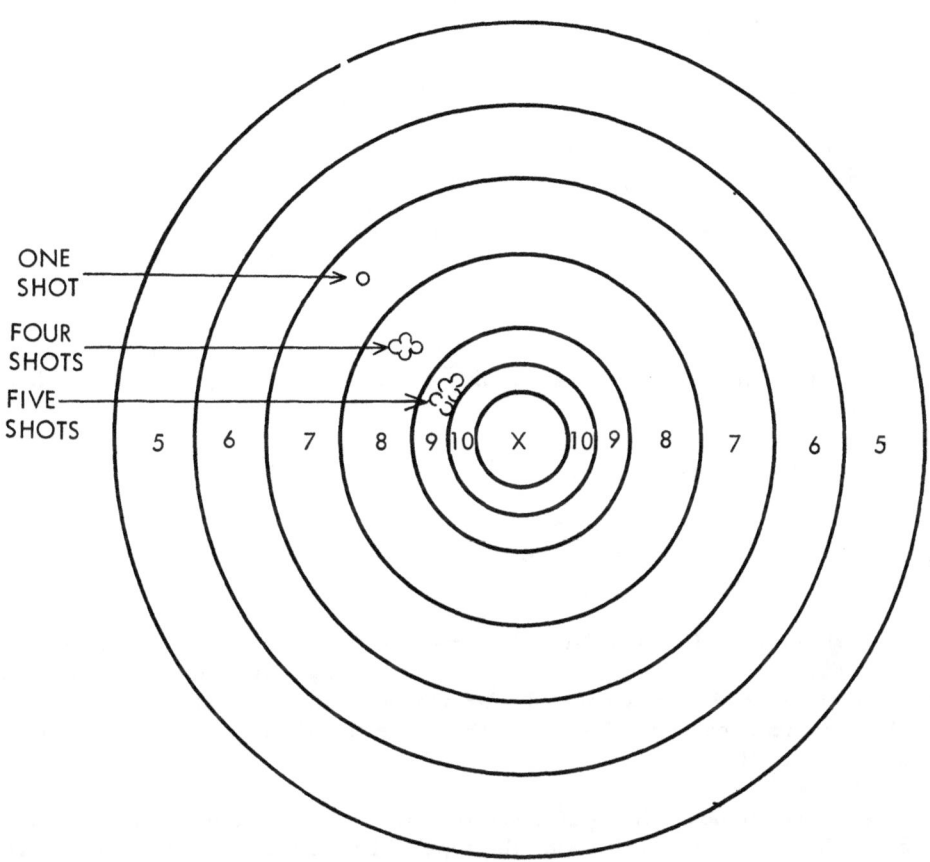

Figure 3. Grouping of Shots.

       a. In case of keyholed or tipped shots the higher value is awarded if the leaded edge of the bullet hole touches the scoring ring of higher value even though the hole is elongated to the bullet's length rather than being a circle of the bullet's diameter.

       b. In case of skid shots the higher value is awarded if the leaded edge of the bullet hole touches the scoring ring of higher value, except the value of a skid shot may not be more than one ring higher than the original point of bullet contact with the target. The target shall be defined as the entire card or paper on which the scoring rings are printed but shall not include the backing. When the original point of bullet contact is outside the target card it will be scored as a miss. When the original point of bullet contact is on the target card but outside the scoring rings and the leaded edge of the bullet touches a scoring ring it will be given the value of the lowest scoring ring.

       c. When a bullet enters a target from the back side it will be scored as a miss.

   4. (14.4) Misses - Hits outside the scoring rings are scored as misses. (Except as provided in 14.3 b).

   5. (14.5) Early or Late Shots - When stationary target frames are used if any shots are fired at the target before the command "Commence Firing" or after the command "Cease Firing" the shots of highest value equal to the number fired in error will be scored as misses.

   6. (14.6) All Shots Count - All shots fired by the competitor after he has taken his position at the firing point will be counted in his score, even if the pistol is accidentally discharged.

   7. (14.7) Hits on Wrong Target - Hits on the wrong target are scored as misses.

   8. (14.8) Ricochets - A hole made by a ricochet bullet does not count as a hit and will be scored as a miss. It must be noted that a bullet which keyholes is not necessarily a ricochet.

   9. (14.9) Visible Hits and Close Groups - As a general rule only those hits which are visible will be scored. An exception will be made in the case where the grouping of three or more shots is so close that it is possible for a required shot or shots to have gone through the enlarged hole without leaving a mark. In this case the shooter will be given the benefit of the doubt and scored a hit.

   10. (14.10) Excessive Hits - If more than the required number of hits appear on the target, any shot which can be identified by the type of bullet hole as having been fired by some competitor other than the competitor assigned to the target or as having been fired in a previous string will be pasted and will not be scored. If more than the required number of hits then remain on the target a complete new score will be fired and the original score will be disregarded except (a) if all hits are of equal value the score will be recorded as the required number of hits of that value. (b) If the competitor wishes to accept a score equal to the required number of hits of lowest value, he shall be allowed to do so. (c) If a competitor fires less than the prescribed number of shots through his own fault, and there should be more hits on the target than the shots fired, he will be scored and number of shots of highest value equal to the number he fired and given a miss for each unfired cartridge. (d) If a competitor, through his mistake, fires more than the required number of shots, he will be scored the required number of hits of lowest value. This shall not be considered a refire as outlined in Rule 9.24.

   11. (14.14) Scorer's Duties -

       a. When targets are scored before removal from the frame the scorer records the value of each hit on score card while holding that card in such position that competitor may see score being recorded. While marking score on card the scorer announces each hit value in an audible tone of voice. Example: "Target Number 2, 2 tens, 2 nines and 1 seven." It is the duty of each competitor to watch the marking of his score on score card and to challenge such scoring immediately if he believes scoring to be incorrect. After each target is scored shot holes are pasted or target is changed.

      b. When targets are scored after removal from target frames, they are removed by target detail on the command, "CHANGE TARGETS" and given to the statistical office for official scoring.

    12. (14.15) Score Cards - Score cards will be prepared by the Statistical Office and delivered to the range officers who will check the target assignments of competitor as he reports at the firing point, then give the score card to the scorer. At the conclusion of each relay range officers will take up the score cards and deliver them to the statistical office.

    13. (14.16) Erasures on Score Cards - Erasures on score cards are not permitted. If correction is necessary, it must be made and initialed by the scorer or range officer. To make correction, the scorer or range officer draws a line, or lines, through the incorrect score and places the correct score above. When targets are scored in pit the record value of any shot will not be changed (except when redisked or re-marked) unless some special message with reference to it is received by the Range Officer from one of the Pit Officers.

    14. (18.14) Signatures - At the conclusion of the score the scorer will add the value of the shots, place the total on the score card and sign the card. The competitor checks value of individual shots, the total, and signs the card. Team Captains verify and sign score cards in team matches. If the competitor or Team Captain desires to protest he shall write "Protested" on score card above his signature.

    15. (10.5) & (18.3.1) Competitors Will Score - Competitors will act as scorers when requested to do so by the Executive Officer or Chief Range Officer, except that no competitor will score his own target.

    16. Decision of ties (RULE 15) -

      a. (15.3) Single Stage - At any range or stage ties will be ranked by applying the following steps, (1) to (6) inclusive, in the order listed below:

        (1) By the greatest number of X's.

        (2) By the fewest misses.

        (3) By the fewest hits of lowest value.

        (4) By the fewest hits of next lowest value, etc.

        (5) In slow fire individual matches by inverse order of shots counting singly from the last shot to the first shot (this will be applied only when targets are being scored after each shot).

        (6) In matches scored in strings of five or 10 shots by the highest ranking score in last string, by highest ranking score in next to last string, etc. (If still a tie apply Rule 15.10.)

      b. (15.4) Multiple Stage - In matches fired in stages, ties will be ranked by applying following steps, (1) to (4) inclusive, in the order listed below:

SCORING (RULE 14)

SINGLE STAGE (NRA PISTOL SCORECARDS) MULTIPLE STAGE

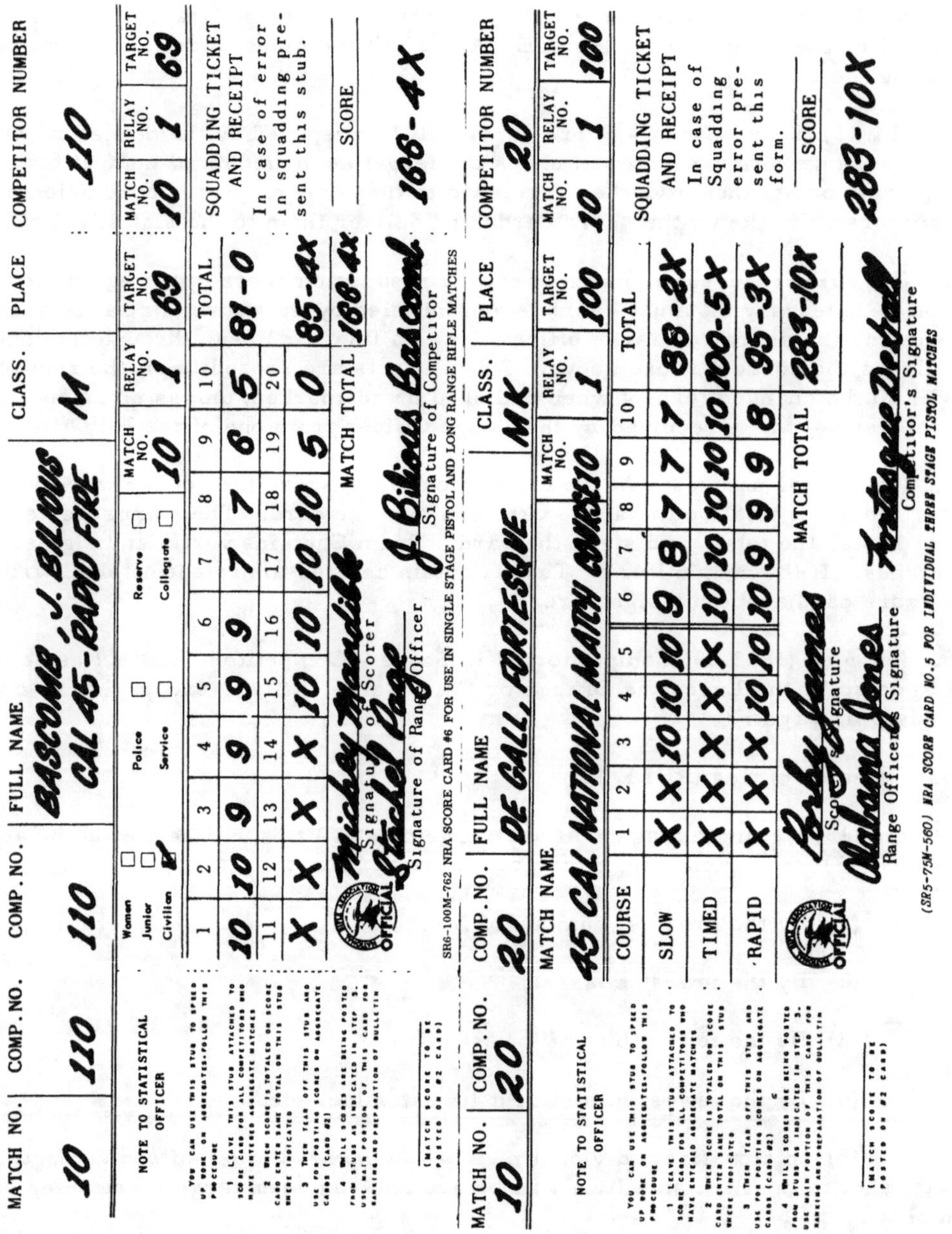

MATCH NO.	COMP. NO.	COMP. NO.	FULL NAME	CLASS.	PLACE	COMPETITOR NUMBER
10	110	110	BASCOMB, J. BILIOUS CAL 45-RAPID FIRE	M		110

NOTE TO STATISTICAL OFFICER

Women ☐ Junior ☐ Civilian ☑ Police ☐ Service ☐ Reserve ☐ Collegiate ☐

MATCH NO.	RELAY NO.	TARGET NO.	MATCH NO.	RELAY NO.	TARGET NO.
10	1	69	10	1	69

1	2	3	4	5	6	7	8	9	10	TOTAL
10	10	9	9	9	9	7	7	6	5	81-0
11	12	13	14	15	16	17	18	19	20	
X	X	X	X	10	10	10	10	5	0	85-4X

MATCH TOTAL 166-4X

Mickey Mandible — Signature of Scorer

Stachel Page — Signature of Range Officer

J. Bilious Bascomb — Signature of Competitor

SQUADDING TICKET AND RECEIPT

In case of error in squadding present this stub.

SCORE 166-4X

(MATCH SCORE TO BE POSTED ON #2 CARD)

SR6-100M-762 NRA SCORE CARD #6 FOR USE IN SINGLE STAGE PISTOL AND LONG RANGE RIFLE MATCHES

MATCH NO.	COMP. NO.	COMP. NO.	FULL NAME	CLASS.	PLACE	COMPETITOR NUMBER
10	20	20	DE GALL, FORTESQUE	MK		20

NOTE TO STATISTICAL OFFICER

MATCH NAME	MATCH NO.	RELAY NO.	TARGET NO.	MATCH NO.	RELAY NO.	TARGET NO.
45 CAL NATIONAL MATCH COURSE	10	1	100	10	1	100

COURSE	1	2	3	4	5	6	7	8	9	10	TOTAL
SLOW	X	X	10	10	10	9	8	7	7	7	88-2X
TIMED	X	X	X	X	X	10	10	10	10	10	100-5X
RAPID	X	X	X	10	10	10	9	9	9	8	95-3X

MATCH TOTAL 283-10X

Porky James — Scorer's Signature

Alabama Jones — Range Officer's Signature

Fortesque DeGall — Competitor's Signature

SQUADDING TICKET AND RECEIPT

In case of Squadding error present this form.

SCORE 283-10X

(MATCH SCORE TO BE POSTED ON #2 CARD)

(SR5-75M-560) NRA SCORE CARD NO.5 FOR INDIVIDUAL THREE STAGE PISTOL MATCHES

Figure 4. Scorecards, Individual.

(1) By the greatest number of X's.

(2) By the highest ranking score at rapid fire; if still a tie, rank each rapid fire score by applying Rule 15.3. If this does not break tie apply Rule 15.4(c).

(3) By the highest ranking score at timed fire; if still a tie, rank each timed fire score by applying Rule 15.3. If this does not break tie apply Rule 15.4(d).

(4) By the highest ranking score at slow fire; if still a tie, rank each slow fire score by applying Rule 15.3. If this does not break the tie, see Rule 15.10.

c. (15.5) Aggregate Matches - In aggregate events ties will be ranked by applying following steps, (1) to (4) inclusive, in the order listed below:

(1) By the greatest number of X's.

(2) By the highest ranking total rapid fire score (including both single and multiple stage match rapid fire scores ranked as shown in Rule 15.3). If this does not break tie apply Rule 15.5(c).

(3) By the highest ranking total timed fire scores (including both single and multiple stage match timed fire scores ranked as shown in Rule 15.3).

(4) By the highest ranking total slow fire score (including both single and multiple stage match slow fire scores as shown in Rule 15.3). If still a tie see Rule 15.10.

d. (15.7). Team Matches - Ties in team matches will be ranked in the order shown below:

(1) By considering team score as though it were a single score fired by an individual. The same precedent applies as that indicated above (15.3 to 15.5 inclusive).

(2) By highest individual aggregate score.

(3) By second highest individual aggregate score, etc.

(4) By highest individual score, second highest individual score, etc., at each stage considered in the order listed in Rule 14.4.

e. (15.10) Unbreakable Ties - In case a tie cannot be ranked under the provisions of the above rules of this section the Executive Officer will direct that the tie be decided and prizes awarded under such one of the following plans as appears necessary or advisable.

(1) By firing of a complete or partial score under the original match conditions or at longest range of the match.

(2) By drawing of lots for merchandise, medal or trophy awards and combining any cash awards to which those tied may be entitled and equal division of such cash among those tied.

f. (9.12) PERFECT SCORE: In individul matches competitors who fire a perfect score will continue to fire five-shot strings as soon as practicable until a hit is made outside the scoring ring of highest value to provide a means of breaking ties and establishing National Records over courses for which National Records are recognized. In three stage matches containing slow, timed and rapid fire the competitor will continue to fire five-shot strings at rapid fire until he has made a hit outside the scoring ring of highest value. When using these scores to break ties, rules will apply as provided in Section 15. This rule does not apply in team matches.

F. **COMPETITOR'S DUTIES AND RESPONSIBILITIES.** (RULE 18)

1. (18.1) Discipline - It is the duty of each competitor to sincerely cooperate with tournament officials in the effort to conduct a safe, efficient tournament. Competitors are expected to promptly call the attention of proper officials to any infraction of rules of safety or good sportsmanship. Failure of a competitor to cooperate in such matters or to give testimony when called upon to do so in any case arising out of infractions of these rules may result in said competitor being considered as an accessory to the offense.

2. (18.2) Knowledge of Programs - It is the competitor's responsibility to be familiar with the program. Officials cannot be held responsible for a competitor's failure to obtain and familiarize himself with the program.

3. (18.3) Eligibility - It is the competitor's duty to enter only those events for which he is eligible and to enter himself in the proper classification.

4. (18.3.1) Competitors Will Score - Competitors will act as scorers when required to do so by the Executive Officer or Chief Range Officer, except that no competitor will score his own target.

5. (18.4) Classification - It is the competitor's duty to have his current Classification Card in his possession when using a classification system. Unclassified competitors must obtain their Score Record Book from the Official Referee, Supervisor, or tournament officials.

6. (18.5) Individual Entries - In individual matches it is the duty of the competitor to make his own entries on the forms and in the manner prescribed for that tournament. Errors due to illegibility or improper filling out of forms are solely the competitor's responsibility. The statistical office is not required to accept corrections after entry closing time.

7. (18.6) Squadding Tickets - It is the competitor's duty to secure his squadding ticket for each match (or to consult the squadding bulletin) in ample time to permit reporting at the proper time and place to fire each match. It is not the duty of officials to page competitors in order to get them on the firing line. Competitors upon receipt of squadding tickets should inspect them for correctness of competitor's number and non-interference in squadding assignments. Errors should be reported immediately to Statistical Officer.

8. (18.7) Reporting at Firing Point - Competitors must report at their assigned firing point immediately when the relay is called by the Range Officer. The proper pistol and ammunition for that particular match must be ready and in safe firing condition. Time will not be allowed for pistol repairs, sight blacking, sight adjustments or search for missing equipment after a relay has been called to the firing line.

9. (18.8) Timing - Time for the firing of a string (within the official time limit) is the competitor's responsibility. Range officers will not announce the time during the firing but if requested will give the competitor information as to remaining time.

10. (18.9) Loading - No competitor will load a pistol except at the firing point and after command has been given by the range officer.

11. (18.10) Cease Firing - Competitors must immediately obey this command whether or not they have finished their string. Even though pressure has been applied to trigger, pressure must be released so that the shot will not be fired. Assume the "ready" position pending further orders.

12. (18.11) Checking Scores - It is the duty of competitors to check their scores as written on score card and to sign score card at the conclusion of match. When scoring is done in the Statistical Office competitors must promptly check Preliminary Bulletin and call attention to errors within the time specified at the tournament. Failure to check scores within time limit forfeits the right to challenge.

13. (18.12) Clearing the Firing Point - It is the competitor's duty to leave firing point promptly at the conclusion of his relay. When leaving the firing point pistols must be unloaded. Cylinders must be open on revolvers, slide locked back and a magazine removed on semi-automatics.

14. (18.13) Checking Bulletin Board - It is the duty of all individual competitors and team captains to check Bulletin Board between each match. The Statistical Officer must be immediately notified of apparent errors. Official Bulletins must be checked and the Statistical Officer notified of any discrepancies between the Preliminary and Official Bulletins. Official Notices on the Bulletin Board have the same effect as conditions printed in program. It is the duty of competitors to familiarize themselves with all such Official Notices.

15. (18.14) Score Cards Must Be Signed - At the conclusion of the score the scorer will add the value of the shots, place the total on the score card and sign the card. The competitor checks value of individual shots, the total and signs the card. Team Captains verify and sign score cards in team matches. If the competitor or Team Captain desires to protest he shall write "Protested" on the score card above his signature.

16. (18.15) Responsibility - It shall be the competitor's responsibility:

    a. That all equipment meets all rules and match specifications in any match in which that equipment is to be used.

    b. That competitor's position conforms to the rules.

    c. That competitor has full knowledge of the rules under which the match is fired.

    d. That after due warning of any infraction of existing rules, that competitor shall understand that a repetition thereof shall be the subject of disqualification for that match or tournament.

G. TEAM OFFICERS' DUTIES AND POSITION. (RULE 12)

1. (12.1) Team Captain - In team matches, other than two-man teams, each team must have a designated Team Captain. He is responsible for maintaining discipline within his team squad. He will at all times cooperate with the officials of the tournament in the interests of safety, efficiency and good sportsmanship. A Team Captain is responsible for all the duties of members of his team. In team matches it is his responsibility to:

    a. Be familiar with the program.

    b. Make proper entries.

    c. Have team members report at proper firing point at the right time, ready to fire.

    d. Check scores, sign score cards and make challenges.

    e. Check Preliminary and Official Bulletin and Official Notices.

f. Make protests.

g. Collect prizes.

TEAM CAPTAIN'S DUTIES (RULE 12)

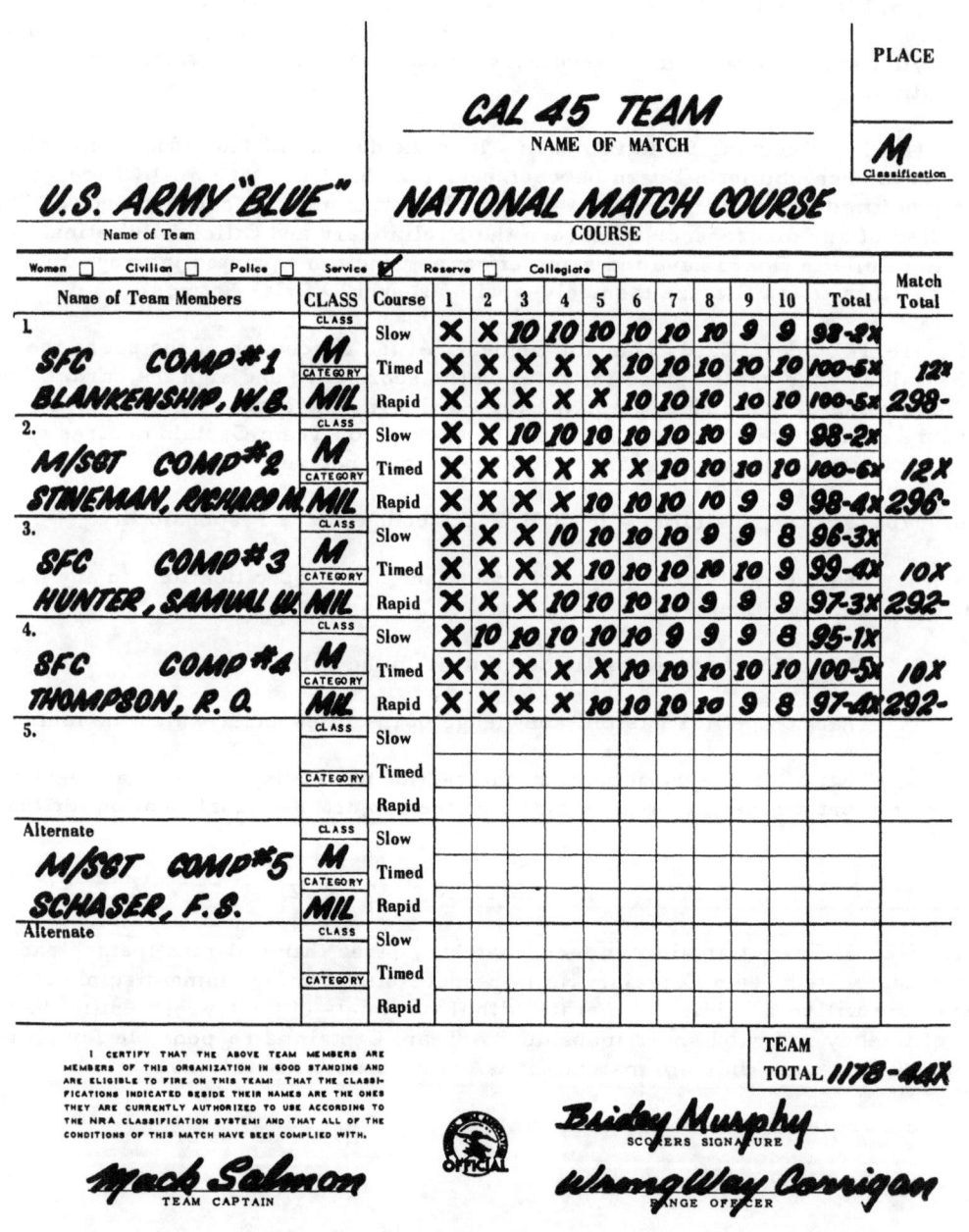

Figure 5. Score Card for Pistol Team Match.

2. (12.2) Team Coach - The Team Coach is the Team Captain's deputy performing such duties as the Captain may assign to him. The Coach serves as Team Captain in the absence of the latter, and under such circumstances becomes responsible for maintaining discipline within the team and for all other responsibilities of the Team Captain.

3. (12.3) Team Entries - In team matches it is the duty of the Team Captain to make entries on the forms and in the manner prescribed for that tournament. All shooting members of the team must be named on the entry form. If the Team Captain or Coach is a firing member of his team he must be so listed on the entry form. No alternates may be used unless alternates are provided for in the match conditions and have been named on the entry form. If alternates are permitted in the match conditions and are to be used, they must be named on the entry form at the time the entry is filed. A shooter may not be listed as an alternate on more than one team in any match. Competitors listed as principals on one team cannot be named as alternates on another team entered in the same match. Errors due to illegibility or improper filing out of forms are solely the Team Captain's responsibility.

4. (12.4) Team Captain and Coach, Position - In team matches the Team Captain or Coach will be allowed on the firing line in such position as not to interfere with the proper operation on the range.

5. (12.5) Team Coach in Team Matches - Coaching is permitted in all team matches within the team only. The coach may assist team members by calling shots, checking time, checking scoring, ordering sight changes, etc.; but he must so control his voice and actions as not to disturb other competitors. He will not physically assist in loading or in making sight corrections. (Example of team score card preceding page.)

H. ELIGIBILITY OF COMPETITORS - INDIVIDUAL AND TEAM. (RULE 2)

1. (2.1) Members of the National Rifle Association - Any individual member in good standing, including Benefactors, Patrons, Endowment, Life, Annual, Associate, Non-Resident and Junior Members.

2. (2.2) Civilian - Any civilian including all members of the Reserve Officers Training Corps (ROTC, NROTC and AFROTC), personnel of the State Security Forces (e.g., State Guard organizations having no federal recognition), retired members of each of the several services comprising the Armed Forces of the United States, and members and former members entitled to receive pay, retirement pay, retainer pay or equivalent pay, are classified as civilians except as noted in the example below. All competitors who are enrolled undergraduates of any of the service academies will be considered as civilians, and may compete in collegiate and ROTC categories. Reserve or National Guard personnel who during the present calendar year have not competed as National Guard, Regular Service or Reserve and have not been provided service support, wholly or in part, may fire as civilians.

EXAMPLE: Members of the regular Army, Navy, Air Force, Marine Corps, Coast Guard; members of the reserve components on active duty; retired personnel of the several services comprising the Armed Forces of the United States on active duty; or police (Rule 2.4) are not permitted to compete as civilians.

3. (2.3) Junior - Any boy or girl who has not reached his or her nineteenth (19) birthday who is either an individual NRA member or a member of an NRA affiliated club. A junior's age at the start of a torunament, league or match series will govern his eligibility.

EXAMPLE: A junior whose 19th birthday will be June 19th enters a tournament, the matches of which will be fired on June 18, 19 and 20. Because he is a junior at the start of this tournament he is eligible to continue the tournament on June 19 and 20.

4. (2.4) Police - Any regular, full time member of a regularly constituted law enforcement agency, including the enforcement officers of the several departments of the United States Government: State, County or Municipal Police Departments; Highway Patrols; Penal Institution Guards; full time salaried Game Wardens, Deputy Game Wardens and Deputy Sheriffs; regularly organized Railroad or Industrial Police Departments, Bank Guards and Armored Truck and Express Company Guards.

Special Officers, Honorary Officers, Civilian Instructors, Deputy Sheriffs, Deputy Game Wardens or Police Officers who are not on a full time, full pay basis in a single department are not eligible to compete as police.

5. (2.5) National Guard - Federal recognized officers or enlisted men of the Army National Guard, the Air National Guard, or the Naval Militia of the several states, territories, the District of Columbia, or the Commonwealth of Puerto Rico, who are not on extended active duty.

6. (2.6) Regular Service - Officers or enlisted men of the Regular United States Army, Navy, Air Force, Marine Corps, Coast Guard, and members of reserve components thereof, who are on extended active duty; provided the team "reserve components" shall include Army National Guard and Air National Guard called into federal service and while in such status.

7. (2.7) Reserve - Officers and enlisted men of any reserve components of the Armed Forces, exclusive of the Army National Guard and the Air National Guard of the United States, not on extended active duty.

8. (2.8) College - Regularly enrolled undergraduate students who comply with the eligibility rules of their institution. An under-graduate is a student who has not received his bachelors degree.

9. (2.9) School - Regularly enrolled undergraduate students of any primary or secondary school, who comply with the eligibility rules of their institution.

10. (2.10) Team Representation - No competitor may fire on more than one team in any one match.

NOTE: Except in double events ("two-man teams") entries will not be accepted from "Pickup" teams (teams whose members are selected without regard to club or other organization affiliation) unless the program specifically provides for such eligibility.

11. (2.13) Regular, National Guard or Reserve Teams - Members of such teams must have been commissioned or enlisted members of their respective service for a continuous period of at least thirty days immediately preceding the day of competition. Army National Guard, Air National Guard, and Naval Militia personnel may be combined into a single team.

12. (2.20) Residence - In those matches which are limited to residents of any specified geographical area a "resident" is defined as:

a. A person whose place of domicile has been within the specified area for at least thirty days immediately prior to the day of the match, whether or not his employment is at a place requring him to commute or travel into some other area.

b. A person who has been regularly employed within the specified area for at least thirty days immediately prior to the day of the match and who has maintained domicile in that area for the same period of time, although his permanent residence is located outside the specified area.

c. Military, Naval and Air Force Personnel: The place of residence of members of the Military, Naval and Air Force establishments on active duty is defined as the place at which they are stationed by reason of official orders, provided they have been so stationed within the specified area for a period of at least thirty days immediately prior to the day of the match. In the case of retired, Reserve, or National Guard personnel not on active duty, the provisions of paragraphs a. and b. will apply. Naval personnel assigned on sea duty qualify for a residence in the area which is the usual base or home port of the unit to which attached.

d. Federal and State Law Enforcement Officers: The provisions of paragraph c. will apply.

CONCLUSION: A competitor who has full knowledge of the pistol match rules and understands range procedures, has relieved his mind of anxiety during and after the shooting phases as to whether any of these numerous essentials have been overlooked. This will allow him to concentrate on controlling his shooting to the fullest extent of his ability.

NOTE: NRA Pistol Match Rules are subject to various, detailed annual changes. Each competitor should obtain a personal copy for study and familiarization with the numerous rules governing the conduct of a registered pistol match.

CHAPTER XV

NRA PISTOL RANGE PROCEDURE AND SAFETY RULES

It is the duty of each competitor to sincerely cooperate with tournament officials in the effort to conduct a safe, efficient tournament. Competitors are expected to promptly call the attention of proper officials to any infraction of rules of safety or good sportsmanship.

The safety of competitors, range personnel and spectators requires a high degree of self-discipline, constant attention to range control commands, the careful handling of firearms, exercising caution in moving about the range area, acting and speaking with consideration toward other competitors.

A. <u>RANGE CONTROL</u> (RULE 10)

1. (10.1) Discipline - The safety of competitors, range personnel and spectators requires continuous attention by all to the careful handling of firearms and caution in moving about the range. Self-discipline is necessary on the part of all. Where such self-discipline is lacking is the duty of range personnel to enforce discipline and the duty of competitors to assist in such enforcement.

2. (10.2) Loud Language - Loud or abusive language will not be permitted. Competitors, scorers, and range officers will limit their conversation directly behind the firing line to official business.

3. (10.3) Delaying a Match - No competitor may delay the start of a match through tardiness in reporting or undue delay in preparing to fire.

4. (10.3.1) Preparation Period - In all cases competitors will be allowed three minutes to take their places at their firing point and prepare to fire after the firing point has been cleared by the preceeding competitor.

5. (10.4) Policing Range - It is the duty of competitors to police the firing points after the completion of each string. The range officer will supervise such policing and see that the firing points are kept clean.

6. (10.5) Competitors Will Score - Competitors will act as scorers when requested to do so by the Executive Officer or Chief Range Officer, except that no competitor will score his own target.

7. (10.6) Repeating Commands - A range officer will repeat the Chief Range Officer's commands only when those commands cannot be clearly heard by competitors under his supervision.

B. <u>RANGE FIRE COMMANDS</u> (RULE 10.7)

1. (10.7) Firing Line Commands - When ready to start firing a match the range officer (usually the Chief Range Officer) commands "RELAY NO. 1, MATCH NO._____ (or naming the match) ON THE FIRING LINE." Each competitor in that relay then immediately takes his assigned place at his firing point and prepares to fire but does not load. If this is a new relay, the range officer will announce "THE PREPARATION PERIOD STARTS NOW." At the end of three (3) minutes, the range officer states "THE PREPARATION PERIOD HAS ENDED," then proceeds with the match.

The range officer having made sure that the range is clear (in timed and rapid fire targets must also be turned out of firing position) then commands "WITH 5 ROUNDS, LOAD."

2. The range officer then asks, "IS THE LINE READY?" Any competitor who is not ready or whose target is not in order will immediately raise his arm and call, "Not ready on target . . . ." The range officer will immediately state, "The line is not ready," and the range officer will immediately investigate the difficulty and assist in correcting it. When the difficulty has been corrected, the range officer calls, "THE LINE IS READY."

3. When the range officer asks, IS THE LINE READY?" and the line is ready, he then calls, "THE LINE IS READY."

4. The range officer then commands, "READY ON THE RIGHT. READY ON THE LEFT." Competitors may point their guns toward the target after the command, "Ready on the right." The range officer will then command, "READY ON THE FIRING LINE." The targets will be exposed or the signal to commence firing will be given in approximately three seconds.

The range officer then commands "COMMENCE FIRING" which means to start firing without delay as timing of the string is started with this command. "Commence Firing" may be signalled verbally, by a short sharp blast on a whistle or by moving the targets into view.

5. "CEASE FIRING" is the command given by the Range Officer at the end of time limit for each string or at any other time he wishes all firing to cease. Firing must cease immediately. Even if a competitor is about to let off a carefully aimed shot he must hold his fire and open the action of his gun. Failure to immediately obey this command is one of the worst infractions of range discipline as it may result in the wounding or death of some man, woman or child who has wandered into the line of fire somewhere on the range or behind the targets. On this command cylinders shall be opened or slides locked back and all guns placed on the shooting stand and not handled until the next command of the Range Officer. "Cease firing" may be signalled verbally, by a short sharp blast on a whistle or by moving the targets out of view.

6. After the command to "cease firing" is given at the end of a string, a subsequent command is given, "CEASE FIRING - UNLOAD - CYLINDERS OPEN - SLIDES BACK - GUNS ON THE TABLE." On this command all assistant range officers and scorers check their competitors to make sure each one obeys the command before signalling the Range Officer that their portion of the firing line is clear.

7. When all assistant range officers and scorers have given the "clear" signal the range officer commands "SCORE TARGETS AND PASTE" (or "CHANGE").

Other commands used less frequently are:

"POLICE FIRING POINTS" means pick up fired cartridge cases, empty cartridge cartons and "tidy-up" the firing line.

"AS YOU WERE" means disregard the command just given. For example, if the commands were given "Ready on the right" followed by "As you were" it would mean someone was not ready

"CARRY-ON" means proceed with whatever was being done before some interruption occurred.

C. PISTOL RANGE SAFETY RULES (DA FM 23-35)

The coach and the shooter must exercise utmost care in all phases of handling weapons and ammunition. Cleaning weapon, dry fire practice, transporting guns and ammuntion to and from quarters or on trips to matches seem to be more dangerous than when the equipment is being used in a match or organized practice. The average person is less apt to commit a dangerous act when in a crowd where many weapons are present. This, no doubt, is why you seldom hear of anyone being wounded or killed at pistol or rifle matches. The constant supervision and caution in handling firearms is well served. A coach or range official would not hesitate to admonish when observing even the slightest infraction.

The following items are listed to enable the shooter to avoid jeopardizing the safety of others or himself:

1. Clear the pistol every time it is picked up for any purpose. Never trust your memory. Consider every pistol as loaded until you have proved it otherwise.

2. Always unload the pistol if it is to be left where someone else may handle it.

3. Always point the pistol up when snapping it after examination. Keep the hammer fully down when the pistol is not loaded.

4. Never place the finger within the trigger guard until you intend to fire or to snap for practice.

5. Never point the pistol at anyone you do not intend to shoot, nor in a direction where an accidental discharge may do harm. On the range, do not snap for practice while standing back of the firing line.

6. Before loading the pistol, draw back the slide and look through the bore to see that it is free from obstruction.

7. On the range, do not insert a loaded magazine until the time for firing.

8. Never turn around at the firing point while you hold a loaded pistol in your hand, because by so doing you may point it at the man firing alongside of you.

9. On the range, do not load the pistol with a cartridge in the chamber until immediate use is anticipated. If there is any delay, lock the pistol and only unlock it while extending the arm to fire. Do not lower the hammer on a loaded cartridge; the pistol is much safer cocked and locked.

10. In reducing a jam first remove the magazine.

11. To remove a cartridge not fired, first remove the magazine, and then extract the cartridge from the chamber by drawing back the slide.

12. Safety devices should be frequently tested. A safety device is a dangerous device if it does not work when expected.

13. Don't mix alcohol with gun powder.

14. Make sure range is clear before firing if you are not participating in organized firing, fire only on command of range officer.

15. Do not handle weapon when any person is forward of firing line.

16. No ammunition to leave range.

17. Do not fire over and above barrier or backstop.

18. Protect weapons from theft, thereby preventing loss from weapons falling into the hands of irresponsible persons. Do not leave weapons unattended either on the range or in a car. Lock all shooting equipment in trunk of car when transporting. Equipment exposed in an unlocked car is a temptation needlessly advertised.

19. Register personal weapons on post, camp, or station.

CHAPTER XVI

NATIONAL TROPHY PISTOL INDIVIDUAL AND TEAM MATCH RULES

NATIONAL TROPHY PISTOL MATCH CONDITIONS. (AR 920-30, Condensed)

1. NATIONAL TROPHY MATCHES:

    a. The National Trophy Matches are conducted by the National Match Director for the National Board for the Promotion of Rifle Practice.

    b. These matches consist of the following events, fired on the dates indicated:

    (1) The National Trophy Individual Pistol Match

    (2) The National Trophy Pistol Team Match

2. MANDATORY PARTICIPATION: Any competitor, military or civilian, whose transportation or other expenses are paid, either partially or wholly, from government funds (appropriated) or quasi-public funds (exchange, recreation, welfare and morale) will be required to:

    a. Enter and complete firing in the appropriate National Trophy Individual Match, unless eliminated under the provisions of paragraph 18 of this program.

    b. Enter and complete firing in the appropriate National Trophy Team Matches as eligible under these regulations.

    (Alternates present and authorized to fire are considered eligible to draw travel pay and per diem.)

    c. The term "complete firing" means that in individual matches the individual must fire in all stages of the match, and in team matches all firing members must fire in all stages of the match (except as noted in match conditions).

3. GENERAL CONDITIONS: The rules and regulations for the conduct of the National Trophy Matches are set forth in AR 920-30/OPNAV INST 3590.7A, AFR No 50-17, dated 13 March, 1962, as amended.

    Except as specifically noted in AR 920-30, in this program or in Match Director's Bulletins, the National Trophy Matches will be conducted under the rules and regulations as prescribed in the latest edition of the Official Rules of the National Rifle Association of America for the conduct of pistol competitions.

4. COMPETITOR'S RESPONSIBILITIES:

    a. It is the competitor's responsibility to know and abide by the provisions of:

    (1) AR 920-30/OPNAV INST 3590.7A, AFR No. 50-17, dated 13 March, 1962, as amended.

    (2) The program of the Tournament being fired.

    (3) NRA rules as applicable.

(4) Appropriate National Match Director's Bulletins.

b. In team matches the team captain is charged with this responsibility for his team.

5. ENTRIES:

a. General--Entries of individuals and teams in the National Trophy Matches must be made with the Statistics Directors for each match, as outlined herein.

b. NO ENTRIES WILL BE ACCEPTED AFTER CLOSING TIME STATED FOR EACH MATCH.

c. Individual--Entry must be made on the regular entry card enclosed with the program. Mail entry to Statistics Director, The National Matches, Camp Perry, Ohio. Entry must be received not later than: 15 July--National Trophy Individual Pistol Match.

d. Changes to Entries--The original entry card should be carefully checked to be sure that all information necessary and the correct information is given. A confirmation card will be sent all competitors, and this should be carefully checked to be sure the information is correct. If there is any change to be made in the entry, do not wait until you get to Camp Perry, but WIRE OR PHONE THE STATISTICS DIRECTOR IMMEDIATELY. Failure to enter appropriate data on your entry card will not constitute grounds for challenge or protest in the case of special awards.

e. Team--Mail entries for teams will not be accepted. Special forms will be issued at Camp Perry to team captains by the Statistics Director upon request. In order to be eligible for team entry all team members must submit individual entry cards as indicated in paragraphs 19, 20, 21 and 22. If entry in NRA Matches or National Trophy Matches is not desired, card should be marked "Team Match Only."

6. ENTRY FEE: No entry fee will be charged for any National Trophy Match.

7. INDIVIDUAL ELIGIBILITY:

a. The National Trophy Match competitions are open to all United States citizens, both male and female, who are at least 16 years of age and to all members of the United States Armed Forces, both active duty and Reserve Components.

b. Certain competitive events and certain awards are restricted to special categories of shooters as a matter of grouping contestants for the best interest of fair competition.

c. Juniors (under 19) who will remain in camp overnight or who wish to draw equipment must be accompanied by an adult NRA member who will sign for their equipment and be responsible for their conduct and welfare.

d. Individuals who are under 16 years of age may not compete in any National Trophy Match.

8. COMPETITOR CATEGORY:

a. In National Trophy Matches, competitors will register and participate in one and only one of the following categories:

Regular Service, Reserve, National Guard, ROTC, Civilian, Police. (Police category for pistol matches only.)

b. The following eligibility requirements will govern in determining proper categorical match participation:

(1) Regular Service Category. Regular Service, Reserve and National Guard personnel on extended active duty (90 days or more) with the Regular Service will participate in this category only.

(2) Reserve Category. Members of any Reserve Branch of the United States Armed Forces (to include all members who are in the standby Reserve), except Army and Air National Guard, Naval Militia, and members of the Fleet Reserve, will participate in this category unless qualified as Regular Service under (1) above, or if a bona-fide member, an individual may select and participate only in Police or ROTC category.

(3) National Guard Category. Members of the Army or Air National Guard or Naval Militia will participate in this category unless qualfied as Regular Service under (1) above, or, if a bona-fide member, an individual may select and participate only in Police or ROTC category.

(4) Police Category. (Pistol Matches only) Fulltime, paid members of any police organization will participate in the Police category, or, if a bona-fide member, an individual may select and participate in Reserve or National Guard category. A policeman nay also participate in the Civilian category if he attends as one of two NBPRP authorized police members of a State Civilian Pistol Team provided he is not a member of a police organization that enters a team in the National Trophy Pistol Team Match. In the National Trophy Individual Pistol Match all policemen, except those who enter as National Guard or Reserve, will participate in the Police category.

(5) ROTC Category. Members of any high school or college ROTC unit (Army, Navy, or Air Force) and Service Academies of the Armed Forces will participate in this category, or, if a bona-fide member, an individual may select and participate in Reserve, National Guard, Police or Civilian category.

(6) Civilian Category.

(a) Individuals who are not in any of the above listed categories will fire as civilians.

(b) Retired Service personnel, including Fleet Reservists, will fire as civilians or, if a bona-fide member, an individual may select and participate only in Police category.

9. AMMUNITION: In the National Trophy Matches service ammunition will be issued by range personnel at the firing line. Competitors are required to fire this ammunition and none other. A competitor will be disqualified if any other ammunition of the same caliber as issued is found about his or her person while in position at the firing line.

10. WEAPONS: Pistol--U.S. Pistol, caliber .45 M1911 or M1911A1 or the same type and caliber of commercially manufactured pistol. The pistol must be equipped with issue or similar factory type standard stocks (i.e., without thumb rests and without tape or other wrappings on the stocks.

Trigger pull must be not less than 4 pounds. The pistol must be equipped with open sights. The front sight must be nonadjustable. The pistol may be equipped with an adjustable rear sight with open U or rectangular notch, the distance between sights measuring not more than 7 inches from the apex of the front sight to the rear face of the rear sight. The forestrap of the receiver grip may be checkered. The main spring housing may be either the flat or arched type, checkered or uncheckered.

Trigger shoes may be used. Trigger stops internal or external are acceptable. Otherwise, external alterations or additions to the arm will not be allowed. The internal parts of the pistol may be specially fitted and include alterations which will improve the functioning and accuracy of the arm providing such alterations in no way interfere with the proper functioning of the safety devices as manufactured. All standard safety features of the weapon must operate properly. It is the competitor's responsibility to have his weapon checked by National Match Ordnance personnel prior to the firing of the match.

11. USE OF WEAPON BY MORE THAN ONE COMPETITOR: No two competitors may fire the same rifle or pistol in any competition comprising the National Trophy Matches.

12. FIRING POSITIONS AUTHORIZED:

    a. Standing. Erect on both feet, no other portion of the body touching the ground or any supporting surface.

    b. In the National Trophy Pistol Match no special consideration will be given to the physically handicapped shooter.

13. LOADING AND RELOADING: By command of the range officer only.

14. FIELD GLASSES AND TELESCOPES: Competitors may use field glasses or telescopes on the firing points unless otherwise prescribed in the conditions of the match.

15. COMPETITOR MUST VERIFY SCORE: The competitor (team captain in team matches) must verify the correctness of the score card, including the name on the card, the value of each individual shot and all other data on the card. The competitor acknowledges the correctness of all data on the card when he signs it. Should a competitor or team captain sign an incorrect card or leave the firing line without signing the score card, no challenge or protest will be allowed. If the competitor or team captain desires to protest he will write the word "Protested" on the score card above his signature.

16. COMPETITORS PRESENT PUNCTUALLY: Competitors will be present at the firing points punctually at the time stated on their squadding tickets. It is the competitor's responsibility to appear at the firing point to which he has been assigned, prepared to fire, when his relay in any match is called to the firing line. Any competitor who fails to report on the proper relay or firing point will forfeit the right to fire in that match unless the competitor presents satisfactory evidence to the National Match Director that he is late through fault of the National Match staff. In team matches, the first pair only need be present at the time set for firing to begin.

17. STATION OF COMPETITORS: Each competitor will remain on or in rear of the assembly line in rear of his firing point until called by the range officer to take his position at the ready line or firing point. No one except the officials of the matches, members of the National Board for the Promotion of Rifle Practice, team officials, the competitors on the ready line and on the firing points, scorers and others on duty will be permitted in front of the assembly line without special permission of the Range Director.

18. ELIMINATION OF TEAMS OR INDIVIDUALS: The National Match Director may, at his discretion and by such standards as he may prescribe, eliminate teams and/or individuals of the lowest standing at any time after the first stage of a team or individual trophy match is completed.

19. TEAMS AUTHORIZED: The following teams are authorized in the National Trophy Team events:

a. Active Service, Reserve and Service Academy Teams--See paragraph 15, AR 920-30

b. National Guard and Naval Militia Teams--One each representing each State, the Commonwealth of Puerto Rico and the District of Columbia. Such teams may be composed of mixed Army National Guard, Air National Guard or Naval Militia personnel.

c. ROTC Teams--One each representing the Reserve Officer's Training Corps (Army, Navy or Air Force) from each of the respective Army Area commands, Naval Districts and Air Force ROTC Liaison Areas within the Continental United States. ROTC students from units in Alaska and Hawaii are authorized to compete as members of the Sixth US Army Area ROTC Rifle Team. ROTC students from Puerto Rico are authorized to compete as members of the Third US Army Area ROTC Rifle Team.

d. Civilian Teams--

(1) One each representing each of the States, the Commonwealth of Puerto Rico and the District of Columbia. State Association teams may compete in lieu of state civilian teams, but may not compete if the state is represented by an approved State civilian team.

(2) One each representing each civilian shooting club organized under the rules of the NBPRP and in good standing on the rolls of the DCM. State Associations teams will not be considered a civilian shooting club.

e. Police Teams in Pistol Team Match--One each representing each regularly organized Federal, State, County or Municipal law enforcement agency in the United States.

20. <u>TEAM MEMBERSHIP REQUIREMENTS:</u>

a. At least one of the firing members of each team will be an individual who has never before fired as a member of any team which has competed in that particular event. Firing members who meet the requirements of this paragraph may be counted against the requirements of paragraphs c and d below.

b. No one who has fired on any team which has placed in the top 15% of competing teams in the same event in any two out of three National Trophy Team Matches immediately proceding may be a firing member or alternate.

c. At least three of the firing members of rifle teams and two of the firing members of pistol teams selected to represent the Armed Forces (Army, Navy, Air Force, Marine Corps and Coast Guard), active, Reserve or National Guard, will be individuals who have never before fired on any team which has placed in the top 15% of teams competing in the same event in past respective rifle or pistol National Trophy Team Matches.

d. Teams representing ROTC, Service Academies, civilian or police organizations, will be required to have at least two firing members on rifle teams and one firing member on pistol teams, who have never before fired on any team which has placed in the top 15% of teams competing in the same event in past respective rifle or pistol National Trophy Team Matches.

e. Participation as a shooting member of a Service Academy team or ROTC team will not be considered as previous participation within the above eligibility requirements provided the team did not place in the leg winning category.

f. No individual may be a member of any of the authorized teams unless he has been a bona-fide member of that particular group or organization from which the team is selected for at least 90 days prior to the opening date of the National Matches except that: Members of State civilian teams will be confined to bona-fide residents of the State which the team represents who have lived in the State for at least 30 days prior to the opening date of the National Matches and are otherwise qualified.

21. ALTERNATES: Not later than the time designated for the closing of entries, each team captain will submit to the Statistics Director at his office, on score cards furnished by the Statistics Director for the purpose, a legible list of members and alternates, INDIVIDUALS MAY BE LISTED AS "TEAM COACH," "FIRING MEMBERS" AND "ALTERNATES" OF ONE TEAM ONLY. A team captain and/or team coach may also serve as firing member or alternate, provided he is so listed on the entry form and is otherwise eligible. Substitution for a firing member may be made at any time before he fires his first shot in a match, provided the individual has been listed on the score card as an alternate. Once a team member has fired a shot, substitution may be made for him only in case of incapacitating injury and upon approval by the National Match Director.

22. TEAM CERTIFICATION: Team captains will present to the Statistics Director lists of their respective team members to include designation of team officials and other members of the team, the correct first name, middle initial, last name, service number and grade of members as applicable, their individual homes addresses, official military mailing addresses if applicable, and proper certification as to the eligibility of the members of their team under these rules and regulations at the time entry in the match is made. Civilian teams will be recognized as representing a particular state only if attested to by the State Adjutant General or President, Vice President or Secretary of the respective State Rifle or Pistol Association, as accredited by the National Rifle Association or under conditions established by the Executive Officer, NBPRP. Such teams must enter as (name of state) State Civilian Team. Other titles will not be accepted for entry. Individual club presidents will attest to the teams representing their organizations. Teams representing law enforcement agencies and service teams of the Armed Forces of the United States will be certified as provided under appropriate regulations governing each such agency or service. ROTC and Service Academy teams will be certified by appropriate military commanders.

23. COACHING:

    a. Individual Matches. No coaching is permitted.

    b. Team Matches. Coaching is permitted in all team matches within the team only; any member of a team (captain, coach, firing member or alternate) may function as coach provided such member fulfills the eligibility requirements set forth below:

    (1) A team captain or coach must be at least 21 years of age on his last birthday prior to the beginning date of the National Matches and must otherwise meet the eligibility rules of the National Matches;

    (2) Officers, warrant officer, or enlisted personnel of the active military services may act as coaches for teams of their respective services and for civilian, ROTC and Service Academy teams;

    (3) Teams representing the National Guard and teams representing other components of the Armed Forces Reserves may be coached only by officers, warrant officers, or enlisted personnel of the particular reserve component concerned;

    (4) ROTC and service academy teams will be coached only by members of the Armed Forces;

(5) A civilian rifle or pistol team may request the National Match Director to assign a coach, and upon application the National Match Director will assign a coach from the officers and enlisted men or resident civilians. Team captains must be provided from individual team sources;

(6) A coach once assigned will not be changed except for cogent reasons and upon approval of the National Match Director.

24. STATION OF TEAM OFFICIALS:

a. DELETED.

b. In pistol team matches one coach will be allowed for each target assigned to the team.

c. The coach cannot shift his position nor shift the position of the competitors of the pair firing for the purpose of forming a wind shield for the firer.

d. The coach must confine himself to the normal position of a coach and his activities to those normally expected of a coach.

e. The coach may assist team members by calling shots, checking time, checking scoring, ordering sight changes, etc., but must control his or her voice and actions so as not to disturb other competitors.

f. The coach will not physically assist in loading or making sight corrections.

g. In team matches the team captain and one assistant may be seated in front of the assembly line but not in advance of a line established three paces in rear of the line of scorers except as provided herein. They may but only if he actually occupies the coaching position on the line.

25. TEAM REPRESENTATIVES IN PITS: Not applicable to pistol.

26. CLUB QUALIFICATION FIRING: In complying with the requirements of paragraphs 11c and f and 15, AR 920-20, in reporting annual qualification firing, club officials may, upon being furnished sufficient evidence by the firer, count the score made by club members in the National Trophy Match courses toward such annual qualification.

27. CHALLENGES: The procedure outlined in Section 16 of the NRA rules for scoring challenges will be modified to the following extent:

a. Rifle Matches--Not applicable.

b. Pistol Matches--Challenges will be decided first by the challenge officer appointed by the Match Director. If not satisfied with the challenge officer's decision the competitor may file a formal written appeal with the Match Director, provided such appeal is filed within two hours of the time the competitor is notified of the challenge officer's decision. If not satisfied with the Match Director's decision the competitor may file a formal written appeal with the NBPRP Appeals Committee, provided such appeal is filed within 5 hours of the time the competitor is notified of the Match Director's decision.

c. Repeated challenges of the same shot, target, or bulletin will not be permitted.

d. The challenge fee in these matches will be $1.00.

e. All challenge money will be placed in the National Match Fund of the NBPRP.

28. PROTEST:

a. A competitor may formally protest:

(1) Any injustice it is felt has been exercised. (This excludes evaluation of a target which may be challenged as outlined elsewhere.)

(2) The conditions under which another competitor has been permitted to fire.

(3) The equipment which another competitor has been permitted to use.

b. A protest must be initiated immediately upon the occurrence of the protested incident. Failure to comply with the following procedure will automatically void the protest:

(1) State the complaint verbally to the Chief Range Officer or the Statistics Director as appropriate. If not satisfied with that decision then--

(2) File a written protest stating all the facts with the National Match Director. This protest must be filed within five hours of the time the Chief Range Officer's or Statistics Director's decision is made known to the competitor. If not satisfied with that decision, then--

(3) File with the NBPRP Appeals Committee a protest in writing stating all facts in the case. This protest must be filed within five hours of the time the National Match Director's deicsion is made known to the competitor.

c. The NBPRP Appeals Committee will consist of those members of the Board, not less than three in number, who are present at the matches at the time of the protest. The decision of this Appeals Committee will be final.

d. Failure to enter appropriate data on entry card and subsequent nonconsideration for special awards will not be considered grounds for protest or challenge.

29. PENALTIES:

a. Any person interfering with a competitor on the firing line or annoying the competitor in any way will be warned to desist. If upon said warning the offense is repeated, such person will be ordered off the range at once.

b. Any competitor or team will be disqualified from competing further in the matches and may be denied any prize won during the current matches when found guilty by the National Match Director of any of the offenses listed below:

(1) Firing under a name other than that under which entered;

(2) Firing twice for the same prize;

(3) Falsifying scores or being an accessory thereto;

(4) Offering a bribe of any kind to any official or other person;

(5) Evading any of the conditions prescribed for the conduct of any match;

(6) Refusing to obey any instructions of the National Match Director, Range Director or Range Officer;

(7) Being guilty of disorderly conduct;

(8) Violating range safety regulations;

(9) Being guilty of any conduct considered by the National Match Director to be discreditable.

CHAPTER XVII

REQUIREMENTS FOR EARNING A DISTINGUISHED PISTOL SHOT BADGE

Effective 1 January 1963

1. The subject badges are awarded by all the major services in the U.S. military establishment, as a symbol of the highest achievement in the achievement in the field of competitive marksmanship.

2. In order to enhance and ensure the continued prestige of these badges and to provide uniform conditions for their award to servicemen, it is agreed that:

a. Award of Excellence-in-Competition badges and credit toward Distinguished badges shall be made on the basis of individual unassisted performance in recognized individual matches.

b. Award of the appropriate Distinguished badge shall be made when an individual has earned a minimum of 30 credit points for excellence in competition in such recognized matches

c. Credit points shall be awarded to the highest scoring 10% of all non-distinguished participants completing the match ranked in order of merit. Fractions of .5 and over will be resolved to the next higher whole number. Smaller fractions will be dropped.

d. Credit points shall be awarded to winning personnel as determined in paragraph c above on the following basis:

To the highest scoring 1/6	10 points
To the next highest scoring 1/3	8 points
To the remaining personnel authorized credit points	6 points.

e. Each individual shall be authorized to compete for credit points in not more than four recognized matches with the service pistol each calendar year.

f. Matches recognized for award of credit points and authorized participation are as follows:

Competitions	Active Army	USAR ARNG	Cadets USMA
Major Command Championships	X	X	X
US Army Championships	X		
Interservice Championships	X	X	X
National Trophy Individual Matches	X	X	X
NRA Regional Championship Matches) OR Command Matches of other Services)	*	X	X

281

g. Courses of fire, weapons and ammunition for matches in which credit awards are authorized shall be in accordance with those prescribed in the current edition of Rules and Regulations for the National Matches.

h. Non-distinguished personnel who hold "leg" credits for Distinguished award, as of 1 January 1963, shall be credited with 10 points for each such leg, not to exceed a total of 20 points.

i. Badges will be provided eligible Army personnel by the CG, USCONARC, in accordance with the provisions of AR 672-5-1.

j. Members of the other Services and civilian competitors who qualify for credit points toward Distinguished award in major command matches will be reported by the CG, USCONARC as follows:

(1) Members of other Services direct to the Service concerned.

(2) Civilians, to include ROTC students and retired military personnel, to the Director of Civilian Marksmanship, Headquarters Department of the Army.

*NOTE: Authorized only when circumstances preclude participation in a Major Command Championship.

INDEX

PAGE

A

Active Desire to Forget	13
Application (Lesson Outline)	45
Articulation (Voice)	57
Actual Equipment Used as a Training Aid	89
Administrative Responsibilities	226
Administrative Check List	237
Ammunition Rules (NRA)	253
Ammunition Rules (Nat'l Trophy)	274
Alternates (Nat'l Trophy Pistol Match)	277

B

Buzz Session	35
Brain Storming (Original Ideas)	35
Body of the Lesson Outline	42
Body Movement in Speaking	62

C

Control of Learning	9
Concepts Acquired by Learning	11
Critical Thinking Ability	11
Conference Method	19
Conference Planning	21
Conducting a Conference	23
Conducting a Demonstration	27
Conducting Student Performance	29
Committee Conference Workshop	31
Conclusion of Lesson	44
Closing Statement	45
Confident Attitude	67
Control of Interest	69
Classroom Administration	78, 84, 86
Check-up Question	80
Charts, Wall (GTA)	93
Chalk Boards	94
Critique Technique	102
Certificate of Training	115
Chart (CTA) Requirements for Pistol Instructor Course	116
Comment Sheet	122
Clothing and Equipment List	234
Check List of Pistol Match Supplies	238
Competitive Regulations	246, 249
Challenges and Protests (NRA)	252
Competitors Duties	262
Cease Firing	262, 269
Clearing the Firing Point	263
Checking Bulletin Board	263
Checking Score on Score Card	263
Competitor Responsibility, (Nat'l Trophy Pistol Match)	272
Competitor Category, (Nat'l Trophy)	273

  PAGE

Coaching (Nat'l Trophy Pistol Match) 277, 278
Challenges and Protests (Nat'l Trophy) 278

                                    D

Discussion Technique 19
Demonstration Method 24
Demonstration, Advantages 25
Demonstration Forms 26
Demonstration Planning 26
Demonstration, Conduct of a 27
Debate 32
Discussion, Panel 33
Doing: Discussion and Practical Work 72
Discussion by Students 73
Decisions of Ties in Scoring 259
Delaying a Pistol Match 268
Distinguished Pistol Shot Badge 281

                                    E

Estimate of the Teaching Situation 38
Elements of the Lesson Objective 38
Errors in Lesson Outlining 45
Effective Speaking 57
Enthusiastic Attitude 67
Examination Technique & Purposes 100
Examination Characteristics 101
Equipment and Ammunition Rules (NRA) 253
Erasures on Score Card 259
Eligibility of Competitors, Individual (NRA) 265
Eligibility of Competitors, Team (NRA) 266
Entries NRA Match 262
Entries Nat'l Trophy Pistol Match 273
Eligibility, Nat'l Trophy Pistol Match 273

                                    F

Facts Acquired in Learning 11
Forgetting 13
Forum (Class Participation) 31
Forceful Speech 60
Facial Expression in Speaking 65
Follow-Up Question 79
Films (movies) Used as Training Aids 90
Film Strips Used as Training Aids 91
Film Slides (35mm) as Training Aids 92
Fire Commands 268

                                    G

Group Instruction 31
Gain Attention 42
Grammar (Speech) 58

PAGE

Gestures (Speech) 63
GTA Charts (Wall or Easel). 93
Glossary of Instructional Terms 105
GTA Chart Requirements for Pistol Instructor Course 116
General Conditions, Nat'l Trophy Pistol Match 272

H

How You Remember 11
How You Forget 12
Humorous Attitude. 68
Hearing as an Interest Factor. 71
Humor 71
Handling of Student Questions 81
Human Relations in Teaching 84

I

Instructor, The Successful 3
Interference Learning 13
Instruction, Methods of 14
Instructional Policy 14
Instructor's Teaching Notes. 47
Instructional Aids 48
Inflection (Speech) 59
Interest Control 69
Interest Affects Learning 69
Interest Factors (Learning Activities) 71
Interest Span 73
Instructional Devices (Teaching Vehicles) 74
Instruction Management 78, 84
Instructor - Student Relationship 84
Implementing the Training Mission 110
Individual Entries in Pistol Match 262

L

Learning Process 5
Learning Steps. 5
Learning, Control of. 9
Lecture Method 15
Lecture Techniques 18
Lesson Planning. 37
Lesson Objective 38
Levels of Student Proficiency 39
Lesson Outline. 42, 125
Lesson Introduction 42
Lesson Tie-In 43
Lesson Conclusion. 44
Lesson Outline Sample 49
Lead-Off Question 79

PAGE

List of Items Needed for Conducting Pistol Instructor Course 113
Loading of Pistol (NRA) 262
Loading and Reloading of Pistol (Nat'l Trophy) 275

M

Motivation 43
Movement of Body in Speaking 62
Management of Instruction 78
Models as Training Aids 89
Malfunction Causes 243
Maintenance of Weapons 243
Marksmanship Unit Shop 244
Misses in Scoring (NRA) 258
Mandatory Participation, Nat'l Trophy Match 272

N

NRA Pistol Match Rules 248
National Trophy Pistol Match Rules 272

O

Organization of the Lesson 40
Over-Head Projectors 92
Organization of a Pistol Team 226

P

Products of Learning 11
Preferences Acquired in Learning 11
Performance Method of Instruction 28
Practice Exercises (Develope Skill) 28
Problem Exercises 28
Panel Discussion 33, 104
Performance Objectives, Student 39
Pronunciation (Voice) 57
Pauses (Speech) 59
Posture (Speaking) 62
Practical Exercises or Work 73
Purpose of Training Aids 87
Practice Score Card 241
Procurement of Equipment 242
Procurement of Weapons 242
Procurement of Ammunition 242
Pistol Match Rules, NRA 248
Protests (NRA) 252
Perfect Score, Firing for Record 261
Position of Team Captain 265
Position of Team Coach 265
Policing Range 268
Penalties (Nat'l Trophy Pistol Match) 279
Protests (Nat'l Trophy Pistol Match) 279

PAGE

R

Remembering, Types of	11
Recollection	11
Recall	11
Recognition	12
Retain Student Attention	45
Rehearsals	48
Rate of Speech	58
Reaction to Student Answers to Questions	81
Rhetorical Question	80
Resolving Difficult Conference Situations	83
Recording, Tape	95
Review of Weak Points	97
Reporting to Firing Point	262
Residence Eligibility	266
Range Procedure (NRA)	268
Range Control	268
Range Fire Commands	268
Range Discipline (NRA)	268
Requirements for Earning the Distinguished Pistol Shot Badge	281

S

Successful Instructor, the	3
Steps in Learning	5
Skills Acquired in Learning	11
Symposium	32
Seminar (Small Group Discussion)	34
Student Proficiency Levels	39
Student Performance Objectives	39
Supporting Material	40
Scope (Lesson Outline)	43
Summarize (Lesson Outline)	45
Sample Lesson Plan	49
Speaking Effectively	57
Speech Improvement	61
Speaking with Body Movement	62
Speaking with Body Posture	62
Speaking with Gestures	63
Speaking Attitude	66
Sincerity in Attitude	66
Seeing as an Interest Factor	72
Span of Interest	73
Sub-Summaries	82
Selection of Training Aids	87
Slides, Film, 35mm	92
Scheduling Course of Instruction	111
Stages of Development of a Shooter	110
Student Comment Sheet	122
Strip Map Example	233
Scorecard, Practice	241
Shot Group Areas of Dispersion	244

287

	PAGE
Scoring Rules (NRA)	254
Scoring Skid Shots	256
Scoring Shot Groups	257
Scoring Misses	258
Scorers' Duties	258
Score Cards, Example Filled Out	260, 264
Signature on Score Card	259
Safety Rules	270

T

Tips to Assist Learning	6
Types of Remembering	11
Training Goals	11
Tastes Acquired in Learning	11
Team Teaching	32
Teaching Notes	47
Teaching Vehicles (Instructional Device)	74
Technique of Asking a Question	80
Transitions	82
Training Aids	87
Training Aid, Purpose	87
Training Aid Selection	87
Types of Training Aids	88
Training Films	90
Tape Recordings	95
Training Aids Used Properly	95
Training Standards	110
Training Schedule, Weekly	240
Types of Competition	248
Ties in Scoring	259
Team Officers Duties and Position	263
Team Captain (NRA)	263
Team Coach (NRA)	265
Team Entries (NRA)	265
Telescopes, (Nat'l Trophy Match)	275
Teams Authorized (Nat'l Trophy)	275
Team Membership Requirements (Nat'l Trophy Pistol Match)	276
Team Certification (Nat'l Trophy Pistol Match)	277

U

| Using Training Aids Properly | 95 |
| Unbreakable Ties | 261 |

V

| Voice Communication | 57 |

PAGE

W

Why You Forget	13
Wall Charts (GTA)	93
Weak Points Review	97
Weekly Training Schedule	240
Weapons Allowed (Nat'l Trophy)	274
Weapons Use By More Than One Competitor	275